Lecture Notes in Chemistry

T0238018

Edited by:

Prof. Dr. Gaston Berthier
Université de Paris

Prof. Dr. Hanns Fischer
Universität Zürich

Prof. Dr. Kenichi Fukui
Kyoto University

Prof. Dr. George G. Hall
University of Nottingham

Prof. Dr. Jürgen Hinze
Universität Bielefeld

Prof. Dr. Joshua Jortner
Tel-Aviv University

Prof. Dr. Werner Kutzelnigg
Universität Bochum

Prof. Dr. Klaus Ruedenberg
Iowa State University

Prof Dr. Jacopo Tomasi
Università di Pisa

Springer

Berlin
Heidelberg
New York
Barcelona
Hong Kong
London
Milan
Paris
Singapore
Tokyo

R. Carbó-Dorca   D. Robert   Ll. Amat
X. Gironés   E. Besalú

Molecular Quantum Similarity in QSAR and Drug Design

Springer

Authors

Ramon Carbó-Dorca
David Robert
Lluís Amat
Xavier Gironés
Emili Besalú

University of Girona
Institute of Computational Chemistry
Campus Montilivi
1071 Girona, Spain

E-mail: quantum@stark.udg.es

Library of Congress Cataloging-in-Publication Data

Molecular quantum similarity in QSAR and drug design / R. Carbo-Dorca ... [et al.].
    p. cm. -- (Lecture notes in chemistry, ISSN 0342-4901 ; 73)
    Includes bibliographical references and index.
    ISBN 3540675817 (softcover : acid-free paper)
    1. QSAR (Biochemistry) 2. Quantum biochemistry. 3. Quantum pharmacology. 4.
Drugs--Design. I. Carbó, Ramón. II. Series.

QP517.S85 M646 2000
572.8'633--dc21

                                                                00-041287

ISSN 0342-4901
ISBN 3-540-67581-7 Springer-Verlag Berlin Heidelberg New York

Typesetting: Camera ready by author
Printed on acid-free paper   SPIN: 10763155   51/3143/du - 543210

Contents

Foreword

The study of the healing power of chemical compounds, present into the known natural active principles and its subsequent use, shall be seen initially as a pseudoscientific procedure, a continuation of the Chinese and Arabian occultism, which was based in the percentage content of fire, air, earth and water, as well as on the associated qualities: hot-cold, humid-dry..., which the matter was supposed to be formed by. The method, possessing roots in Hippocrates, Dioscorides and Galen was studied, described and polished by Avicenna, Averroës, Bacon and Villanova in the Middle Age. It also was appearing as a study constant during the renaissance and after. On the other hand, the birth of chemistry as a scientific offspring from alchemy propitiated alternative ways of knowledge in order to solve the same problem. In this manner, in the past century, approximately hundred years from now, Sylvester proposed the first molecular description in numerical discrete form, employing ideas which even in present times can be associated within the so called molecular topology. Sylvester's topological model can be considered the seed allowing the origin of this big tree, which is now known as theoretical chemistry.

During all the past time from the first topological model of Sylvester up to now, the proliferation of numerical parameters to describe molecular structures has not ceased to grow larger. Some of these parameters have played a very important role for the understanding of the organic molecules behavior and, by extension, for the comprehension and evaluation of their physical as well as biological properties. In the mind of every specialist are the Hammett's σ, the Taft constants or the octanol-water partition coefficient. Other numerical parameters, such as those derived from the modern topological molecular representation are in a process of constant revision and growing. Thus, the Hosoya and Randic indices, or the Kier's connectivities, among several not so well known numerical data are usual reference descriptors. They are put at the researchers' disposition, and are easily deducible from any molecular representation in form of ordered sets of numerical figures. All of them are profusely studied and employed in present times. The main idea consists into the use of these numerical data in order to obtain information on the molecular trends to possess or acquire certain properties and, even better than this, to determine in which degree or intensity molecules present everything.

At the turn of the almost past century, in a broadly manner first, in a refined way after the twenties, just at the dawn of the thirties, quantum mechanics was precluding the possibility to mathematically describe the molecular structures in the best of complete and precise ways. Quantum theory, based in a physico-mathematical formalism as well as in experimental evidences about the composition and behavior of matter, aroused the hope to find out the definitive procedure for the a priori theoretical evaluation of the molecular properties. In both, the quantum mechanical Schrödinger's description of the hydrogen atom as well as in the relativistic Dirac formalism, there is the seed of the algorithms to find out the definitive manner to study new molecular entities, which even did not to have to be synthesized, in order to analyze theoretically their structure and properties.

Suddenly, the application of quantum mechanics to study molecular structure got an unprecedented help from the first born electronic computers. Their use, thirty years after the Schrödinger works, became a road nevermore abandoned. It is obvious that, after the first computer models offered by UNIVAC and IBM, the future of the solution of Schrödinger's equation was inexorably tied up to numerical procedures, in a great majority already described in the nineteenth century, others waiting to be developed in parallel along the theoretical development of applied quantum mechanics. Everything was connected to the computational needs related to the search for wavefunctions, constituting the solutions of Schrödinger's equation, also called wave equation in remembrance of its origins. According to the quantum mechanical postulates, all the information about a given molecular system in particular, about a quantum object in general, is contained within the so-called system's density (probability) function. The density function is nothing but the squared module of the system's wavefunction, followed by a manipulation to obtain a drastic reduction of the number of implicit function variables. This means that, in order to obtain any information about the molecular properties, even if the molecule is simply constituted by a virtual formula, that information can be obtained from the associated density function. From here the molecular characteristics could be obtained with ease, applying anew quantum mechanical ideas based in turn in well-established theoretical statistical definitions.

In this way, the medieval dream can be considered fulfilled, several centuries after, by means of the conjunction of a physico-mathematical theory and the technology connected with electronic computing. However, to determine the molecular properties, physico-chemical or biological, even in our days presents several problems not exempt of difficulties, whose roots are nearby located into the zone of the unsolvable. In present times, the quantum mechanical parameters are often employed in the same fashion as those coming from topological origins, which were previously referred, without any deeper connection with the quantum mechanical theory than to have its source on Schrödinger equation and wavefunctions. This way to act has an easy explanation in the fact that the wave equation solution and the density function computation were seen, along the computational development of quantum chemistry, as dissociated processes, independent one from another. The appearance of alternative points of view in the nearby time period, culminating with the 1998 Nobel prize award, allow to think into the leading role the density function has in connection with the Schrödinger equation itself, may be throughout the so called Hohenberg-Kohn theorem or by some extension or variant of it.

It will be certainly early to affirm that the relationship between the molecular structure, their properties and quantum mechanics are completely established and that all details and problems are solved. For the moment it seems out of question that the computational process, associated to the theory, which can be called the Quantum Quantitative Structure-Properties Relationships, it is not related to a unique molecule, but at least in the biological activity environment, has to agglutinate the information simultaneously connected to several molecular structures. From here, the quantum mechanical connection between structure and properties can be unequivocally established by means of the quantum similarity theory, an extension of the quantum theory, developed from a first paper published in 1980. The molecular similarity of quantum origin description is founded in the mathematical structure, which contains the density functions as objects of study, and in extracting the information content from the density function, which it is related in turn with the similarities, which are present within the structures made by microscopic particles. This constitutes an important question, which has been already mentioned as a fundamental element of quantum theory. In practice, the so-called quantum molecular similarity consists in the simple comparison of various density functions, associated to two or more molecules, with as much precision as possible. Such comparison is realized by means of numerical calculations, related with the mathematical concept of measure. The best approximate definition of this comparison process result can be based in the fact that a measure can be considered a sort of generalized volume.

From this starting point can be easily deduced that the equations related with structure-properties relationships can be directly connected with the quantum postulates, in a pure fashion, without additional suppositions. The fashionable empirical structure-properties equations more than a century alive, acquire a legitimization right in this manner and they can be considered from now on as more or less coarse approximations of the theoretical quantum procedure. More than this, from the quantum theory of structure-properties relationships it can be deduced that some empirical parameters of experimental origin, like the octanol-water partition coefficient, shall be strongly correlated with the quantum self-similarity measures: those comparisons where only a density function of a unique molecule intervene. For example, due that it is possible to consider as a self-similarity the quantum Coulomb repulsion potential between the electrons of a molecule, at the same time this molecular property can be considered a substitute of the usual empirical parameters, within molecular series with some degree of homogeneity. Another interesting example, for instance, may be related with the substitution of the classical topological matrices connected to some molecular structure. Such matrices can be refined by means of intramolecular quantum similarity matrices constructed by comparison the atomic electronic densities or these belonging to the polyatomic fragments, which form the molecular structure. Search for pharmacophores can be performed in this last case by means of processes connected to quantum mechanics.

The quantum similarity measures involving one or several density functions cannot be considered uniquely as another step to legitimate previous empirical processes. They constitute a manner, not influenced by the research operator, to find out molecular parameters acting as descriptors, unbiased ordered numerical arrays, obtained without any previous supposition but quantum theoretical considerations. From this initial step, a quantum structure-property relationship can be obtained for any known molecular collection. The quantum similarity measures constitute a basis so general for this specific purpose that they can be used to any microscopic system, from atoms and nuclei up to proteins. In fact the well known atomic periodic table can be deduced as a consequence of the application on the atomic electronic structure of the quantum similarity theory. Of course, quantum similarity permits to obtain a relationship between quantum objects and any associated property, according to the spirit linked to the subjacent quantum theory. It is possible, as a consequence, if one wishes to, to establish molecular periodic tables where, in the same way as in the homologous atomic one, the relationships between the implied molecules permit to obtain information about the behavior of new structures. It is important to repeat now, once more, that also it is feasible to deduce, by using the basic postulates of the quantum mechanical theory only, the form, which have to take the quantum structure-property relationships, and it can be said that this form coincides with the linear appearance employed up to present time within the empirical environment. But this is not the unique similarity between both methodologies.

It must be said, finally, that the evolution of the quantum similarity theory and the numerical tests performed along the past twenty years on a large number of varied molecular systems confirm unequivocally that the theory is correct and completely general. The models deduced from the new theory of the quantum structure-property relationships, by means of the quantum molecular similarity, are comparable in precision, if not better than the empirical results. However, the quantum models differ from the empirical ones in the fact that there is no need to obtain them, any other manner to operate based on any other suppositions, than those most usual among all the possible ones, related with the logical structure of applied quantum mechanics.

All these topics constitute the basis of the present book, and will be developed in detail along the different chapters, illustrating each discussion with real-life application examples, most of them unpublished. Finally, we want to gratefully acknowledge professors Robert Ponec and Paul G. Mezey for their enlightening comments and lively discussions. This work has been partially supported by the CICYT grant SAF-96-0158 and the European Commission contract ENV4-CT97-0508. Thanks also to the *Fundació María Francisca de Roviralta* for financial support.

Girona, January 2000

1 Introduction

Molecular similarity attempts to give a quantitative answer to the question: *how similar are two molecules?* It is clear that this is an interesting problem, and that it has no unique answer. The possible solutions will be associated to the type of molecular aspect that one wants to analyze. Due to the fact that molecules are objects ruled by the laws of quantum mechanics, it seems that one of the satisfactory answers to the question ought to be found within this specific discipline. Following this line of thought, the first quantitative measure of the similarity between two molecules, based on quantum-mechanical basic elements, was formulated by Carbó in 1980 [1]. Carbó proposed that a numerical comparative measure between two molecules could be derived from the superposed volume between their respective electronic distributions. This original definition still holds, and constitutes the fundamental tool of the present work. The seminal idea was developed by this author and collaborators [2-6], and the present state-of-the-art can be obtained from various review articles [7-10]. These papers deepen in the quantum-mechanical nature of the definition, connect it with several subjects of chemical and mathematical interest and show a broad amount of possible applications.

Other academic groups pursued the research line initiated by Carbó, with notable contributions to the definition and application of molecular similarity measures. Among these individuals, it must be highlighted the work of J. Cioslowski [11-14], N.L. Allan and D.L. Cooper [15-21] and the group of W.G. Richards [22-25]. From a different perspective, the problem of the definition of a comparative measure between two molecules was posed by W.C. Herndon [26], which substituted the quantum-mechanical magnitudes by elements of graph theory, in a kind of synthesis between topology and molecular similarity. P.G. Mezey [27-31] and R. Ponec [32-36] proposals to molecular similarity measures also include molecular shape and topological features.

Molecular quantum similarity theory has been employed in a large set of topics: as an indicator of the chiral form of a given molecular species [37]; as a functional for finding optimized molecular alignments [38]; as an interpretative tool for the study of chemical reactions [39]; to compare different theoretical calculation methodologies [40]; for assessing the quality of a given basis set [41] and as molecular descriptors to build quantitative structure-activity relationships [42-53], which constitute the main application issue of the present book. Furthermore, as it has been previously commented, although molecules are the main field of application, the definition of quantum similarity measures is general enough to encompass the comparison between other kinds of quantum objects. In fact, similarity between atoms [54,55], atomic nuclei [56,57], intracule and extracule densities [58] and molecular fragments [48,49] has been already described. Finally, it has been proposed a connection between molecular topology and quantum theory by means of the definition of novel indices inspired by quantum similarity concepts [44].

1.1
Origins and evolution of QSAR

A branch of Chemistry of a great interest nowadays is computer-aided drug design. The possibility of designing compounds with well-defined properties while avoiding the expensive costs of experimental synthesis has led to a great effort in basic research. The fundamentals for an effective design are the so-called quantitative structure-activity relationships (QSAR), a discipline which has become rationalized and systematized very recently. QSAR techniques assume that a relationship between the properties of a molecule and its structure exists, and tries to establish simple mathematical relationships to describe –and later, to predict– a given property for a set of compounds, usually belonging to the same chemical family. QSAR analysis encompasses both the definition of molecular descriptors able to characterize satisfactorily different molecular sets and the statistical treatment which can be applied to these descriptors in order to improve their predictive capacity. The importance of this subject has led to the apparition of specialized journals (*Quantitative Structure-Activity Relationships*; *Journal of Computer-Aided Molecular Design*; *Journal of Molecular Modelling*; *SAR and QSAR in Environmental Research*; etc.), as well as monographic volumes and international conferences.

The origin of QSAR techniques can be dated in the past century, when in 1863, Cros, from the university of Strasbourg, observed that the toxicity of alcohols to mammalians augmented when their solubility in water decreased [59]. Crum-Brown and Fraser postulated in 1868 that a relationship between the physiological activities and chemical structures existed [60]. Later, Richet proposed that toxicity of some alcohols and ethers were inversely proportional to their water solubility [61]. Around 1900, Meyer and Overton, independently, established linear relationships between the narcotic action of some organic compounds and a distribution coefficient of the solubility in water and in lipids, describing a parameter that can be considered some precursor of the current log P, the octanol-water partition coefficient [62,63]. In 1939, Ferguson studied the behavior of diverse properties (water solubility, partition, capillarity, and vapor pressure) in relation to the toxic activity of different homogenous series of compounds [64]. Even if these procedures could be established as the roots of the current QSARs, in the late 30's Hammett proposed the first methodological issue, provided with a general scope. Hammett verified that the ionization equilibrium constants of the *meta* and *para* substituted benzoic acids were related. This existing relationship led to the definition of the so-called Hammett σ constant [65,66]. This parameter became a descriptor able to characterize the activity of many molecular sets. Using this approach as an initial step, other descriptors were proposed [67], but lacking of the relevance of Hammett constant.

In 1964, Free and Wilson postulated that for a series of similar compounds, differing one to another by the presence of certain substituents, the contribution of these substituents to the biological activity was additive and depended only on the type and position of the substituent [68]. The Free-Wilson model, however, cannot be applied to molecules whose substituents are not linear combinations of those existing in the training set.

The systematization of QSAR analyses has to be associated to the work of Hansch and Fujita appeared in 1964 [69]. The basis of the Hansch-Fujita model is the assumption that the observed biological activity is the result of the contribution of different factors, which behave in an independent manner. Each activity contribution is represented by a structural descriptor, and the biological activity of a set of compounds is adjusted to a multilinear model. The descriptors most used in the early QSAR analyses are the aforementioned octanol/water partition coefficient (log P), the Hammett σ constant acting as an electronic effect descriptor and the lipophilicity parameter π, defined by analogy to the electronic descriptor. Together with the previously discussed empirical descriptors, the classical models employ other physico-chemical properties as parameters, some of them derived from quantum chemical calculations, namely: partial charges, HOMO/LUMO energies, etc. In those cases where the structure-activity was too complex to be characterized with these descriptors, even other parameters had been and are used, namely binary indicator variables, which take binary digits discrete values according to the presence/absence of certain substituents [70].

Another interesting perspective to the structure-activity relationship problem has been based on molecular topology concepts. This subject, mainly developed by Wiener [71], Kier and Hall [72] and Randic [73], represents numerically the topological features of the molecules through the so-called connectivity and distance indices. These topological indices have also been successfully applied to QSAR [74,75].

In 1988, QSAR techniques suffered a great transformation due to the introduction of the so-called three-dimensional molecular parameters, which accounted for the influence of different conformers, steroisomeres or enantiomeres. This type of models, usually known as 3D QSAR models, also imply the alignment of molecular structures according to a common pharmacophore, derived from the knowledge of the drug-receptor interaction. The first published model possessing these characteristics was the *Comparative Molecular Field Analysis* (CoMFA), proposed by Cramer et al [76], which is currently one of the most widely employed QSAR methodologies. Other different 3D QSAR approaches have been proposed since CoMFA appearance [77-81], some of them associated to concepts of similarity between different molecular aspects.

1.2
Molecular similarity in QSAR

Once QSAR techniques were well established, molecular similarity was also considered as a valid tool to construct prediction models. The underlying assumption for the application of molecular similarity in QSAR is that *similar molecules should possess similar properties*. The different ways to define similarity between molecules lead to the different existing approaches.

The first QSAR papers employing similarity ideas used the similarity between electrostatic distributions to derive QSAR parameters [82,83]. Starting from concepts based on graph theory, Rum and Herndon built a similarity index matrix whose elements were comprised within zero and one. The columns of this matrix were then used as a descriptors in a multilinear regression [84].

A.C. Good and co-workers described a protocol for the application of similarity matrices to QSAR quite similar to that currently in use [85-87]. Similarity matrices were built using electrostatic potentials and shape descriptors. For a first time, treatment of similarity matrices included dimensionality reduction and a statistical validation process.

The application of Quantum Similarity to QSAR was initially made in a qualitative way, trying to associate the spatial grouping of molecules with the value of some of their physico-chemical properties [88,89]. A few years later, a connection between the expectation value of a quantum operator, representing a physical observable and the molecular quantum similarity measures was described. The practical implementation of those ideas has led to the publication of several papers [42-53], and finally, to the present work.

1.3
Scope and contents of the book

This contribution pretends to present an up-to-date revision of Quantum Similarity concepts and their application to QSAR. The role of quantum similarity measures can be summarized on their capacity for being the vehicle producing N-dimensional mathematical representations of molecular structures. The elementary basis of Quantum Similarity framework is given in chapter 2. There, a formal definition of quantum objects is given, which requires the introduction of tagged set and vector semispace concepts, as well as density functions, whose central role in Quantum Mechanics is retrieved in this formalism. Thus, in this scheme, the quantum object concept appears to be inseparably connected to density functions. Then, the general form of molecular quantum similarity measures is introduced, and the concrete definitions for practical implementations are specified. Transformations of these measures, called quantum similarity indices, are also given. Two other important topics related to the application of quantum similarity measures are discussed: first, the Atomic Shell Approximation (ASA), a method for fitting first-order molecular density functions for a fast and efficient calculation of the quantum similarity measures. Afterwards, two possible solutions to the problem of molecular alignment, a determinant procedure in all 3D QSAR methodologies.

In chapter 3, the application of Quantum Similarity to QSAR is discussed in detail. The theoretical connection between Quantum Similarity and QSAR, via the discretization of the expectation value law in Quantum Mechanics is shown. Quantum similarity measures are unbiased molecular descriptors, because their values are not chosen according to *a priori* designs: they are built up as a consequence of the theoretical quantum framework results and only depend on the nature of the studied molecular set. Multilinear regressions, as the set of algorithms for building predictive models, are described, together with the statistical parameters to assess the goodness-of-fit and to validate the models. Particularities and limitations of QSAR models based on quantum similarity measures are outlined in this chapter.

Chapters 4 to 7 show different possible approximations to the QSAR problem using quantum similarity measures and provide several application examples. The first approach (chapter 4) uses the entire quantum similarity matrix to derive the descriptors. The convenient pretreatment is discussed in full, including dimensionality reduction and variable selection. Three application examples are given, encompassing three environments of chemical interest: medicinal chemistry, molecular toxicity and protein engineering. Another interesting approach arises when using only the diagonal terms of the similarity matrix, yielding a one-parameter model based on quantum self-similarity measures (chapter 5). Self-similarities are first used to correlate 2D classical descriptors such as log P and Hammett σ constant, and then a direct application to QSAR is shown. Another one-parameter QSAR model can be constructed using the electron-electron repulsion energy as a descriptor. In chapter 6, a formal connection between this descriptor and Quantum Similarity is discussed, where it is proved that the mathematical expression of the electron-electron repulsion energy can be reinterpreted as a special kind of quantum self-similarity measure. Its use as a predictive parameter is illustrated with several examples. Finally, chapter 7 shows one of the possible extensions of Quantum Similarity to other non-molecular quantum objects: the application of the formalism to construct comparative measures between atomic nuclei, and the association of these measures with physical properties of interest.

2 Quantum objects, density functions and molecular quantum similarity measures

In this chapter, the elementary basis of Quantum Similarity framework is presented in an elementary way. Here, by defining in a rigorous way the concept of quantum object, the quantum mechanical concept of Quantum Similarity is described. This leads to a discussion about the role of density functions in the chemical description of molecular structures. Some definitions –tagged, Boolean and functional tagged sets, as well as vector semispaces– are previously introduced, in order to produce the adequate formalism from where Quantum Similarity can be easily deduced and after this computational algorithms can be developed.

2.1
Tagged sets and molecular description

Consider a collection of objects of arbitrary nature constituting the set $\mathscr{S}$, the *object set*, and a collection of mathematical elements, such as Boolean strings, column or row vectors, matrices, functions, etc. that form another set $\mathscr{T}$, the *tag set*. The latter can be generally considered to be completely independent from the former. Both sets can be related by means of the construction of a new composite set $\mathscr{Z}$, a *tagged set*, accordingly to the following definition:

Definition 2.1. Let us suppose a known given object set, $\mathscr{S}$, and a set made up of some mathematical elements, called tags, forming a tag set $\mathscr{T}$. A *tagged set*, $\mathscr{Z}$, can be formed by the ordered product $\mathscr{Z} = \mathscr{S} \times \mathscr{T}$:

$$\mathscr{Z} = \left\{ \forall \theta \in \mathscr{Z} \mid \exists s \in \mathscr{S} \wedge \exists t \in \mathscr{T} \rightarrow \theta = (s,t) \right\} \tag{2.1}$$

Tagged sets are mathematical structures present in the chemical information body. A molecular structure, as a chemical object of study, has a large, steadily growing amount of attached attributes. Such a situation can be generalized by setting a tagged set formal construction rule, where molecules become elements of the object set and their ordered features can be associated to the tag set part.

2.1.1
Boolean tagged sets

Tagged sets can be considered as sets that join simultaneously their elements and any kind of coherent information describing or somehow attached to them. Let us suppose that in a known tagged set, the tag set part is made up of elements expressed in terms of N-dimensional Boolean or bit strings. Any situation has therefore a translation into the integer sequence $\{0,1,2,...,2^{N-1}\}$. Furthermore, every bit string can be associated to any of the 2^N vertices of an N-dimensional unit length hypercube, $\mathbf{H}_N$, with one vertex placed at the origin of coordinates, $\mathbf{0}=(0,0,...,0)$, and the farthest one at position $\mathbf{1}=(1,1,...,1)$. The rest are located in the positive hyperquadrant at intermediate distances. Thus, any set of objects can be structured as a tagged set simply by using the appropriate dimension hypercube vertices as the tag set.

It is also obvious that other tagged sets can be transformed into a Boolean form: consider a tag set made by *N-tuples* of rational numbers as a straightforward example. The nature of the molecular information precludes the possibility of easily transform chemical tagged sets into Boolean molecular tagged sets. In this sense, Boolean tagged sets can be considered as a sort of canonical form, able to describe any kind of rational information attached to a given object set.

2.1.2
Functional tagged sets

Up to now, tagged sets have been supposed to be implicitly constructed by N-dimensional vector-like tag sets. Nevertheless, there is no need to circumscribe tag set parts to any finite-dimensional space subsets. As a previous step, let us suppose that a ∞-dimensional hypercube is taken as the tag set part. A parallelism between the ∞-dimensional vertices and the elements of the [0,1] segment then arises. Furthermore, it must be considered the possible use of Boolean matrices of arbitrary dimension as candidates to the tag set part.

Tag sets can also be made of elements coming from a ∞-dimensional function space. Among all the possible function families, the most appealing candidate corresponds to a subset of some probability density functional space. There are two important reasons for considering this: a) probability density functions are normalizable; b) according to the interpretation proposed by von Neumann [89] and Bohm [90], probability density functions derived from the squared modulus of state wavefunctions are the meaningful mathematical elements, which can be attached to the descriptive behavior of quantum systems [91,92]. This magnitude will be defined in a following section.

In this sense, probability density function tagged sets seem to be the natural ∞-dimensional extension of the discrete N-dimensional Boolean tagged sets. No redefinition is necessary for this purpose: tagged set definition 2.1 still holds when any ∞-dimensional space subset is employed as the tag set part.

2.1.3
Vector semispaces

Due to the great flexibility one has when defining a particular molecular set, it can be useful examining which type of tags can be employed to fill the existing gap between the Boolean hypercubes and the probability density function spaces. A suitable proposal may be associated to N-dimensional vector spaces with some appropriate restrictions. The concept of *vector semispace* arises naturally from this scheme.

Definition 2.2. A *vector semispace* (VSS) over the positive real field $\mathbf{R}^+$ is a vector space with the vector sum part provided by a structure of abelian semigroup.

By an additive semigroup [93] it is understood here to be a group without the presence of reciprocal elements, and hence no subtraction is available. Going further in the previous picture, all the VSS elements can be viewed as being directed towards the region of the positive axis hyperquadrant. Throughout this contribution will be assumed that null elements are included both in the scalar field as well as in the VSS structure. Operations within the VSS structure, such as linear combinations, must be made using positive definite coefficients, in order to maintain the derived elements within the semispace structure. Metric VSS can be constructed in the same manner as the common metric vector spaces, but remembering that all the implied inner products are positive definite, and as a result, no negative cosines of vector angles can be obtained.

2.2
Density functions

A microscopic system is subdued to different force fields, which in Quantum Mechanics are represented as different contributions to a mathematical operator, the Hamiltonian. Once the Hamilton operator is constructed, the time-independent Schrödinger's equation can be formulated, which relates the Hamiltonian to the state energies by means of the wavefunction of the system:

$$\mathscr{H}\,\Psi = \mathscr{E}\,\Psi \tag{2.2}$$

For a N particle system in a given state, the associated wavefunction Ψ is a function of a $2N$-dimensional vector, which depends on the spatial coordinates of the particles and on their spins. The wavefunction is an element belonging to a Hilbert space defined over the complex field, $\mathscr{H}(\mathbf{C})$.

From this wavefunction, a state *density function* ρ can be evaluated by computing the squared modulus of the wavefunction:

$$\rho = \Psi^+ \Psi = |\Psi|^2 \tag{2.3}$$

The density function also depends on all the spatial coordinates and spins. The previous definition is called the *generating rule*, $\mathscr{R}(\Psi \to \rho)$, and it can be rewritten as:

$$\mathscr{R}(\Psi \to \rho) = \left\{ \forall \Psi \in \mathscr{H}(\mathbf{C}) \to \exists \rho = \Psi^+ \Psi = |\Psi|^2 \in \mathscr{H}(\mathbf{R}^+) \right\} \tag{2.4}$$

In summary, the quantum description of a microscopic system is made by the following three-step procedure:

- Construction of the Hamiltonian operator, $\mathscr{H}$
- Computation of the state energy-wavefunction pair $\{\mathscr{E}, \Psi\}$, by solving the Schrödinger equation $\mathscr{H}\Psi = \mathscr{E}\Psi$
- Evaluation of the state density function, $\rho = |\Psi|^2$

It is well known how the dimensionality of the density function can be reduced. An rth-order density function can be defined by integrating the raw eq. 2.3 over all the spins and the entire coordinates except r of them [90]. As a useful particular case, the integration over all the N spins and $N-1$ coordinates yields the *first-order density function*:

$$\rho^{(1)}(\mathbf{r}) = \int \cdots \int |\Psi(\mathbf{r}_1, \mathbf{r}_2, \ldots, \mathbf{r}_N; s_1, s_2, \ldots, s_N)|^2 \, d\mathbf{r}_2 \ldots d\mathbf{r}_N \, ds_1 \ldots ds_N \tag{2.5}$$

These functions are those that will be used for later definitions, as well as in all the applications of this book. In order to avoid unnecessary repetitions, first-order density functions will be referred simply as "density functions", and from now on, they will be written without any supraindex.

Density functions can be interpreted as the probability of finding a particle of the system at a given position $\mathbf{r}$. This probabilistic picture induces a normalization such as:

$$\int \rho(\mathbf{r}) d\mathbf{r} = 1 \tag{2.6}$$

Thus, the integral over the entire space yields the total probability of presence: one. Density functions belong to the real positive definite semispace, $\mathscr{H}(\mathbf{R}^+)$.

Any density function can be obtained by manipulation of the squared modulus of some system wavefunction. Nevertheless, this is not the only possible transformation available. It has been recently described a novel methodology to extend Hilbert vector spaces. In this picture, the wavefunction is associated to a positive definite hermitic operator. An analogue generating rule can be defined, where the derived density function adopts a different functional structure [94]. However, this is not the objective of this book, and the extended densities will not be further discussed.

2.3
Quantum objects

The idea of a quantum object without a well-designed definition appears frequently in the literature. The previous definitions of tagged set, vector semispace and density function entail the background mathematical structure necessary to introduce the quantum object concept.

Definition 2.3. A *quantum object* can be understood as an element of a tagged set: quantum systems in well-defined states are taken as the object set part, and the corresponding density functions constitute the tag set part.

As observed, density functions are conferred a leading role in the quantum mechanical description, which will be reinforced by later explanations. Definition 2.3 is an attempt to provide a rigorous mathematical formalism to a widely employed concept in theoretical chemistry. A deeper discussion on these topics can be found in recent contributions [95-97].

2.4
Expectation values in Quantum Mechanics

One of the quantum mechanical postulates states that all the observables ω of a quantum system can be derived from the associated wavefunction by means of the equation:

$$\omega = \langle \Psi | \Omega | \Psi \rangle = \int \Psi^*(\mathbf{r})\Omega(\mathbf{r})\Psi(\mathbf{r})d\mathbf{r} \qquad (2.7)$$

This equation is often referred to as the expectation value of the operator Ω. However, it is not a real expectation value in the statistical sense. An equation defining the expectation value of an operator needs to necessarily involve the density function of the system:

$$\omega = \int \Omega(\mathbf{r})\rho(\mathbf{r})\,d\mathbf{r} \qquad (2.8)$$

Thus, each physical observable is associated to a non-differential hermitic operator Ω. This operator, when acting over the density function, yields the expectation values of the observable. It must be noted that the previous expression can be interpreted as a scalar product or as a linear functional:

$$\omega = \langle \Omega | \rho \rangle, \qquad (2.9)$$

defined within the space where both $\Omega(\mathbf{r})$ and $\rho(\mathbf{r})$ operators belong.

Eq. 2.8 assesses the leading role of the density functions in Quantum Mechanics, and it constitute the basis to construct the fundamental equation for the structure-activity relationships based on quantum similarity measures, which will be discussed later on.

2.5
Molecular Quantum Similarity

As proposed at the beginning of this book, Molecular Quantum Similarity tries to give an answer to a well-known question in chemistry: how similar are two molecules? Defining a similarity measure between two molecular structures is a crucial problem in theoretical chemistry. In general, the roots of Molecular Quantum Similarity are buried into the most fundamental microscopic theory: Quantum Mechanics. Thus, similarity measures shall be naturally based on well-defined quantum mechanical descriptors, the molecular density functions. According to the quantum mechanical postulates, all the information that can be extracted from a quantum system is contained in its density function [89,90]. Hence molecular electron densities are a suitable source for similarity descriptors.

2.6
General definition of the molecular quantum similarity measures (MQSM)

Let D be the tag set of a quantum object set. A MQSM is an application of the direct product $D \otimes D$ into the positive definite real space, $\mathbf{R}^+$, by means of an operator Ω, acting as a weight:

$$\theta(\Omega): D \otimes D \rightarrow \mathbf{R}^+ \qquad (2.10)$$

In practice, this mathematical application can be translated into an integral measure involving the density functions of two quantum objects:

$$Z_{AB}(\Omega) = \iint \rho_A(\mathbf{r}_1)\Omega(\mathbf{r}_1,\mathbf{r}_2)\rho_B(\mathbf{r}_2)d\mathbf{r}_1 d\mathbf{r}_2 , \tag{2.11}$$

where $\rho_A(\mathbf{r})$ and $\rho_B(\mathbf{r})$ are the first-order density functions associated to molecules A and B, respectively; and Ω is a positive definite operator. Thus, by construction, the MQSM are always real and positive definite: $Z_{AB}(\Omega) \in \mathbf{R}^+$. This measure can also be viewed as a *scalar product* between two continuous mathematical functions, in a particular metric:

$$Z_{AB}(\Omega) = \langle \rho_A | \rho_B \rangle \tag{2.12}$$

Once all the MQSM are calculated for each molecular pair, the overall measures can be collected into a matrix: the *quantum similarity matrix* $\mathbf{Z}=\{Z_{AB}\}$.

The exact form of the positive definite operator Ω defines the MQSM type. This is a degree of freedom that the researcher can arbitrarily set, but historically only two operators have been frequently employed in Quantum Similarity studies. These operators constitute the overlap-like and the Coulomb-like MQSM, which will be defined in the following subsections. Other operators may be chosen, but they will not be discussed here [98-100].

2.6.1
Overlap MQSM

The easiest and most used operator for the MQSM is the Dirac delta distribution [101], $\Omega=\delta(\mathbf{r}_1-\mathbf{r}_2)$. This operator yields the *overlap MQSM*, which has the explicit form:

$$\begin{aligned} Z_{AB} &= \iint \rho_A(\mathbf{r}_1)\delta(\mathbf{r}_1 - \mathbf{r}_2)\rho_B(\mathbf{r}_2)d\mathbf{r}_1 d\mathbf{r}_2 \\ &= \int \rho_A(\mathbf{r})\rho_B(\mathbf{r})d\mathbf{r} \end{aligned} \tag{2.13}$$

The second equality uses the properties of the Dirac delta distribution. This type of MQSM is simply the quantitative measure of the superposition of two molecular density functions. Furthermore, this MQSM is related to the spin-spin contact correction [102], a term of the relativistic Breit Hamiltonian [103].

2.6.2
Coulomb MQSM

Another widely employed operator in Quantum Similarity studies is the Coulomb operator $\Omega=|\mathbf{r}_1-\mathbf{r}_2|^{-1}$, which yields a *Coulomb MQSM*:

$$Z_{AB} = \iint \rho_A(\mathbf{r}_1) |\mathbf{r}_1 - \mathbf{r}_2|^{-1} \rho_B(\mathbf{r}_2) d\mathbf{r}_1 d\mathbf{r}_2 \qquad (2.14)$$

Here, the Coulomb operator acts as a weight for the density functions overlapping. Considering the molecular density function as an electron distribution in space, this expression is anything but the extension of the Coulomb law for continuous charge distributions, and therefore, it can be considered, in some sense, as an electrostatic descriptor.

2.7
Quantum self-similarity measures

When two identical quantum objects are compared, the resulting MQSM is known as a *quantum self-similarity measure* (QS-SM). In the present point of view, self-similarities can be considered as the square of the norm of the density functions in the chosen metric:

$$Z_{AA}(\Omega) = \iint \rho_A(\mathbf{r}_1) \Omega(\mathbf{r}_1, \mathbf{r}_2) \rho_A(\mathbf{r}_2) d\mathbf{r}_1 d\mathbf{r}_2 = \|\rho_A\|^2 \qquad (2.15)$$

Self-similarities are the diagonal elements of the quantum similarity matrices. This type of MQSM has a particular relevance and utility, which will be discussed afterwards.

2.8
MQSM as discrete matrix representations of the quantum objects

A MQSM collection can be considered as discrete matrix representations of a given quantum object, which depend on the rest of quantum object tags taken into account. The matrix representation of a known quantum object, with respect to a given quantum object set, can be associated with a set of positive definite numerical values, which constitute a finite-dimensional vector, representing the quantum object.

Let us suppose the studied quantum object set to be made up of n individuals. MQSM can be viewed as n-dimensional projections of the original ∞-dimensional density matrix. As it will be discussed, there exist some transformations that can reduce the dimensionality of the data, and so m-dimensional vectors ($m<n$) can represent the quantum objects, loosing a minimal amount of information.

Given a quantum object s belonging to a quantum object set: $Z=S\times D$, there exists a one-to-one correspondence between this object and the associated density function tag:

$$\forall s \in S \to \exists \rho \in D \Rightarrow s \leftrightarrow \rho \qquad (2.16)$$

The MQSM can be considered as defined in the tensorial product space $D \otimes D$, and collected in a quantum similarity matrix:

$$\mathbf{Z} = \{Z_{AB}(\Omega)\} \qquad (2.17)$$

The matrix $\mathbf{Z}$ contains, therefore, all the similarity relationships between the overall set. The columns of such a matrix:

$$\mathbf{Z} = (\mathbf{z}_1, \mathbf{z}_2, ..., \mathbf{z}_n) \qquad (2.18)$$

can be interpreted as the discrete vector representation of each element of the quantum object set Z. Once fixed the set S and the similarity operator Ω, a unique association between an object $s \in S$ (or a density function ρ) and a $\mathbf{z}$ vector can be established:

$$\forall s \in S (\forall \rho \in D) \to \exists \mathbf{z} \in \mathbf{Z} \Rightarrow s(\rho) \leftrightarrow \mathbf{z} \qquad (2.19)$$

Each object s is seen, from this perspective, as a point in the n-dimensional similarity space where $\mathbf{z}$ belongs: a *point-object*, or in the chemistry framework, as a *point-molecule*. The collection of all the points of the set, described in this way, is known as a *molecular point-cloud*.

2.9
Molecular quantum similarity indices (MQSI)

As it has been previously discussed, once the quantum object set and the weight operator are defined, the MQSM become unique. Nevertheless, the similarity matrix elements can be transformed to obtain normalized or scaled values. These transformations yield the so-called *Molecular Quantum Similarity Indices* (MQSI). The importance of the MQSI leans upon the fact that a normalized measure can be more easily interpreted. In addition, distance-like or correlation-like measures are well-known elements in Multivariate Analysis, for which several treatment techniques have been proposed.

Many MQSI can be defined from the original MQSM [98,99], but only one will be presented here: the cosine-like or Carbó indices. This is the unique MQSI which will be used in the following application examples. In fact, MQSI do not bring forward new information on the molecular similarity relationships, so the election of any one of them should not be decisive for the derived results.

2.9.1
The Carbó index

A normalization of the MQSM can be obtained simply by dividing the original measures by the product of the square root of the corresponding self-similarities:

$$C_{AB} = Z_{AB}(Z_{AA}Z_{BB})^{-1/2} \qquad (2.20)$$

Carbó indices are in the range comprised between zero and one: $C_{AB} \in [0,1]$. Closer to one, more similar will be the compared quantum objects. Carbó indices are correlation-like parameters, and they can be interpreted as the cosine of the subtended angle of the two density functions in ∞-dimensional space:

$$C_{AB} = \frac{\rho_A \mid \rho_B}{\|\rho_A\| \|\rho_B\|} = \cos \alpha_{AB} \qquad (2.21)$$

As in the former MQSM, the Carbó indices can be collected as a matrix: $C = \{C_{AB}\}$. This is the only MQSI which will be used in this work, but nothing opposes to define different ones and use them exactly in the same way as the original MQSM or Carbó indices. Several alternative MQSI definitions can be found in the literature [98,99].

2.10
The Atomic Shell Approximation (ASA)

Nowadays, the *ab initio* theoretical calculation of large molecular systems or transition metal complexes is strongly constrained by the number and type of involved atoms, since the computational requirements grow with the number of basis set functions in the same fashion as LCAO MO molecular computations are. When applied to Quantum Similarity, the main constraining factor in *ab initio* methods is the calculation of four-center integrals appearing in the MQSM expression. These integrals can be readily evaluated if fitted molecular density functions are employed, instead of *ab initio* ones. Different fitting algorithms for the first-order density functions can be found in the literature [104-110], but not all of them fulfill the necessary conditions to guarantee the positive definite structure of density functions.

In this section, a theoretical model of density fitting will be presented: the so-called *Atomic Shell Approximation* (ASA), developed in our laboratory [107-110]. This methodology has been employed throughout the application examples of this book in order to obtain the electronic densities used as a source of MQSM.

2.10.1
Promolecular ASA

ASA consists of expressing the first-order density function of a molecule A as a linear combination of spherical 1S functions provided with positive definite coefficients:

$$\rho_A^{ASA}(\mathbf{r}) = \sum_{i \in A} c_i s_i(\mathbf{r}). \qquad (2.22)$$

The set of coefficients of the expansion is calculated by minimizing a *quadratic error integral function* between the *ab initio* and ASA density functions. The quadratic error integral function is defined with the usual form:

$$\varepsilon^{(2)} = \int \left| \rho_A(\mathbf{r}) - \rho_A^{ASA}(\mathbf{r}) \right|^2 d\mathbf{r}. \qquad (2.23)$$

The minimization is carried out while keeping positive definite the weight coefficients. This ensures that the positive definite structure of the molecular density functions is guaranteed. This guaranty is strictly necessary if the fitted function has to everywhere behave as a probability density function.

A simple but very interesting particular case of the previous development consists of using a *promolecular approximation* [111-113], in which the molecule is defined as a sum of atomic contributions. In the *promolecular ASA*, the molecular density function is simply a sum of atomic densities:

$$\rho_A^{ASA}(\mathbf{r}) = \sum_{a \in A} Z_a \rho_a^{ASA}(\mathbf{r}), \qquad (2.24)$$

where Z_a is the atomic number of each atom a of the molecule A. The atomic density functions are fitted to an *ab initio* calculation, instead of adjusting the whole molecular density. This allows constructing in a very easy way the first-order density function for any molecule, simply by conveniently adding the fitted atomic density functions, stored in a database, at the corresponding atomic positions.

Assuming that ρ_a is normalized to one, the ASA molecular density function is normalized to the total number of electrons of the molecule (N_A):

$$\int \rho_A^{ASA}(\mathbf{r}) d\mathbf{r} = N_A. \qquad (2.25)$$

Each atomic density function is taken as a combination of squared spherical 1S functions, centered on atom a:

$$\rho_a^{ASA}(\mathbf{r}) = \sum_{i \in a} w_i \left| g_i(\mathbf{r} - \mathbf{r}_a; \zeta_i) \right|^2 . \tag{2.26}$$

w_i are the positive definite coefficients that need to be calculated, together with ζ_i, the exponents of the 1S functions, which have been chosen here of gaussian form:

$$g_i(\mathbf{r} - \mathbf{r}_a; \zeta_i) = \left(\frac{2\zeta_i}{\pi} \right)^{3/4} \exp\left[-\zeta_i (\mathbf{r} - \mathbf{r}_a)^2 \right]. \tag{2.27}$$

The number of gaussian functions used to describe the different atoms depends on their size. Other function types, such as STO, can also be used in the same manner, but they will not be discussed here.

2.10.2
ASA parameters optimization procedure

A possible procedure to optimize the ASA parameters basically possesses three stages: (a) generation of ASA exponents using an even-tempered geometrical series [114-116]; (b) calculation of the positive definite coefficients of the basis functions; and (c) refinement of the exponents of the basis functions. The weight coefficients are optimized using the elementary Jacobi rotation technique [117,118]. The 1S gaussian exponents are optimized with a Newton method [119], which requires the knowledge of the analytical form of the gradient vector and the Hessian matrix of the quadratic error integral function.

As it has been already commented, the atomic density functions need to be computed only once, and afterwards they constitute databases for the construction of molecular ASA densities, following expression (2.24). Several adjustments of the atomic density functions have been proposed. The database used in all the application examples of this book comes from a fitting from an *ab initio* calculation using a 3-21G basis [109] for atoms from H to Kr.

2.10.3
Example of ASA fitting: adjustment to *ab initio* atomic densities computed using a 6-311G basis set

To illustrate the application of the method, an example of atomic ASA fitting is presented for an atomic set from hydrogen to argon. This study is an extension of previous works, in which exponents and coefficients were computed for three basis sets: 3-21G basis set for atoms H to Kr [109], a Huzinaga basis set for atoms H to Rn [110] and a 6-21G basis set for atoms H to Ar [120]. Here, atomic density functions for atoms H to Ar, *ab initio* calculated using a 6-311G basis set [121,122] were fitted to a combination of 1S gaussian functions. The adjustment was carried out by minimizing the quadratic error integral function, using the procedure described in references 109, 110. Table 2.1 gathers the main results of the ASA fitting: the adjustment quadratic error, $\varepsilon^{(2)}$, the relative error in the computation of the atomic quantum self-similarity Z_{aa} and the relative error in the computation of one-electron potential energy, $V(r)$. As derived from these indicators, the fitting is very accurate. The set of ASA parameters can be stored in a database for use in Quantum Similarity calculations [123].

2.10.4
Descriptive capacity of ASA

ASA density functions describe with high accuracy the molecular density shape at a low computational cost. A comparative study between molecular densities derived from *ab initio* calculations and promolecular ASA fitting has been recently reported [124]. The analysis of the isodensity contours revealed some of the weaknesses of the method. The main inconvenient of promolecular ASA densities is the high concentration of atomic density at atomic nuclei, leading to a poor description of the atomic bond formation. Figure 2.1 shows the isodensity contours of 2,4,6-trinitrophenol obtained from an *ab initio* calculation with a 6-311G basis and from ASA.

Table 2.1. Fitting results for 6-311 basis set for atoms H to Ar.

		No. Fitted Atomic Functions			
		3	4	5	6
H	$\varepsilon^{(2)}$	1.55E-05	2.17E-06	3.76E-07	1.21E-07
	% Z_{aa}	-0.487	-0.207	0.014	0.007
	%V(r)	-0.217	-0.168	0.026	0.023
He	$\varepsilon^{(2)}$	1.25E-04	1.75E-05	8.92E-07	5.41E-07
	% Z_{aa}	-0.581	-0.261	-0.021	-0.014
	%V(r)	-0.289	-0.230	-0.022	-0.024
Li	$\varepsilon^{(2)}$	1.13E-03	1.14E-04	1.39E-05	2.68E-06
	% Z_{aa}	0.166	0.005	-0.007	0.005
	%V(r)	0.223	0.103	0.048	0.007
Be	$\varepsilon^{(2)}$	1.35E-03	1.21E-04	2.21E-05	1.38E-05
	% Z_{aa}	0.190	0.037	0.034	0.023
	%V(r)	0.572	0.191	0.107	0.097
B	$\varepsilon^{(2)}$	1.91E-03	2.04E-04	4.18E-05	1.45E-05
	% Z_{aa}	0.082	-0.070	-0.068	0.002
	%V(r)	0.111	-0.267	-0.340	0.049
C	$\varepsilon^{(2)}$	2.38E-03	3.06E-04	9.84E-05	1.94E-05
	% Z_{aa}	-0.001	-0.139	-0.133	-0.007
	%V(r)	-0.109	-0.446	-0.507	0.046

Table 2.1. (continued).

		No. Fitted Atomic Functions			
		3	4	5	6
N	$\varepsilon^{(2)}$	2.96E-03	4.76E-04	2.08E-04	2.41E-05
	$\% Z_{aa}$	-0.127	-0.233	-0.216	0.001
	$\%V(r)$	-0.317	-0.624	-0.663	0.008
O	$\varepsilon^{(2)}$	3.74E-03	7.77E-04	2.45E-04	3.09E-05
	$\% Z_{aa}$	-0.291	-0.354	-0.029	-0.011
	$\%V(r)$	-0.566	-0.826	-0.106	0.003
F	$\varepsilon^{(2)}$	4.71E-03	1.20E-03	2.79E-04	3.59E-05
	$\% Z_{aa}$	-0.474	-0.491	-0.028	-0.017
	$\%V(r)$	-0.780	-1.012	-0.094	-0.028
Ne	$\varepsilon^{(2)}$	5.87E-03	1.77E-03	3.17E-04	9.36E-05
	$\% Z_{aa}$	-0.658	-0.633	-0.027	-0.008
	$\%V(r)$	-0.956	-1.149	-0.072	-0.024
Na	$\varepsilon^{(2)}$	8.50E-03	3.50E-03	5.33E-04	1.87E-04
	$\% Z_{aa}$	-1.238	-1.114	-0.096	-0.089
	$\%V(r)$	-2.730	-2.811	-0.785	-0.729
Mg	$\varepsilon^{(2)}$	1.33E-02	4.44E-03	6.76E-04	2.68E-04
	$\% Z_{aa}$	-1.926	0.04	-0.037	-0.036
	$\%V(r)$	-4.288	-0.691	-0.493	-0.412
Al	$\varepsilon^{(2)}$	2.06E-02	4.80E-03	7.18E-04	2.82E-04
	$\% Z_{aa}$	-2.580	0.118	-0.017	-0.019
	$\%V(r)$	-5.743	-0.228	-0.258	-0.261
Si	$\varepsilon^{(2)}$	3.09E-02	5.02E-03	7.10E-04	2.59E-04
	$\% Z_{aa}$	-3.165	0.143	0.001	-0.015
	$\%V(r)$	-6.902	0.016	-0.148	-0.172
P	$\varepsilon^{(2)}$	4.44E-02	5.19E-03	6.82E-04	2.28E-04
	$\% Z_{aa}$	-3.634	0.132	-0.005	-0.016
	$\%V(r)$	-7.796	0.101	-0.114	-0.147
S	$\varepsilon^{(2)}$	6.10E-02	5.35E-03	6.64E-04	2.25E-04
	$\% Z_{aa}$	-3.945	0.120	-0.005	-0.018
	$\%V(r)$	-8.460	0.125	-0.105	-0.133
Cl	$\varepsilon^{(2)}$	7.88E-02	5.50E-03	6.45E-04	1.83E-04
	$\% Z_{aa}$	2.043	0.114	-0.007	-0.02
	$\%V(r)$	1.084	0.122	-0.109	-0.149
Ar	$\varepsilon^{(2)}$	8.24E-02	5.66E-03	6.26E-04	1.64E-04
	$\% Z_{aa}$	2.022	0.111	-0.013	-0.012
	$\%V(r)$	1.302	0.135	-0.095	-0.140

Density (au)	*ab initio* 6-311G density	ASA fitted density
0.05		
0.15		
0.30		

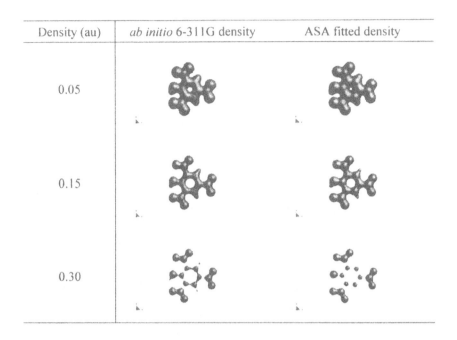

Figure 2.1. Isodensity contours of 2,4,6-trinitrophenol from an *ab initio* calculation with a 6-311G basis set and fitted with ASA. Three density levels are analyzed: 0.05, 0.15 and 0.30 au.

Nevertheless, the final objective of the fitted densities in this work is the fast and accurate calculation of MQSM. Thus, what is really relevant to compare is the differences between the MQSM derived from *ab initio* and ASA densities. With respect to this question, in all the studied cases the differences between the *ab initio* and ASA MQSM were not higher than 2% [109,110]. This allows concluding that the utilization of promolecular ASA densities in a Quantum Similarity computational environment is clearly justified.

2.11
The molecular alignment problem

In this final section of this chapter, it is faced the problem of the alignment between two molecules. As in other 3D QSAR approaches, the overlaying of the involved molecular structures determine the derived results. The determination of the optimal molecular alignment is a problem with application to different branches of Chemistry, for example: 3D QSAR techniques and pharmacophore search [125], the quantitative comparison of the molecular stereochemistry or measures of distortion in crystals [126] or the pattern recognition problem in 3D structural databases [127]. In some 3D QSAR models, the alignment between the molecules determine the derived results, and the main sources of error may be attached both to the molecular conformation and alignment. Two different approaches, namely the maximization of MQSM and the search of a maximal common molecular substructure are discussed.

2.11.1
Dependence of MQSM with the relative orientation between two molecules

All the definitions of quantitative similarity between two molecules involving 3D descriptors depend on the relative position in space between the molecular structures. In particular, MQSM as defined in eq. (2.11) are sensitive to the relative position of both involved molecules. The associated MQSM integral can be rewritten as:

$$Z_{AB}(\Omega; \Theta) = \iint \rho_A(\mathbf{r}_1) \Omega(\mathbf{r}_1, \mathbf{r}_2) \rho_B(\mathbf{r}_2; \Theta) d\mathbf{r}_1 d\mathbf{r}_2 , \qquad (2.27)$$

where Θ symbolizes the relative orientation between the molecules A and B. Thus, a simple, clear and unbiased criterion is needed to align each pair of compared compounds. Unfortunately, there is not a unique way to define a superposition criterion, and different approaches have been proposed so far, as will be discussed below.

When applying Quantum Similarity techniques, two different alignment procedures will be mainly used: one involving the maximization of the MQSM and a second one, which searches for the maximal common substructure of the two molecules.

2.11.2
Maximal similarity superposition algorithm

A possible intuitive alignment criterion when dealing with MQSM can be stated as follows: *two molecules will be optimally aligned when the MQSM integral attains a maximal value.* Thus, one can write,

$$Z_{AB}(\Omega; \Theta) = \max_{\Theta} \iint \rho_A(\mathbf{r}_1) \Omega(\mathbf{r}_1, \mathbf{r}_2) \rho_B(\mathbf{r}_2; \Theta) d\mathbf{r}_1 d\mathbf{r}_2 , \qquad (2.29)$$

which could be named as the *maximal similarity rule*.

In order to obtain the superposition between each molecular pair that makes maximal the similarity integral, an elegant algorithm was designed not long ago [128] and will be briefly described. Let A and B be the two molecules compared, and take A as fixed. Molecule B will be oriented, through translations and rotations, in such a way that eq. (2.29) is satisfied. Let $\{a,a',a''\}$ and $\{b,b',b''\}$ be any three atom collection of molecules A and B, respectively. The first step superposes exactly the pair (*ab*) through the adequate translation of *B*. Then, *B* is oriented in such a way that the axes defined by the segments *aa'* and *bb'* coincide. Finally, the atom *b''* is rotated until the planes defined by the atomic triads are exactly superposed. This process is repeated for all the possible triads of atoms of both molecules, storing the value that maximizes the MQSM. The algorithm includes some restrictions which limit the number of atoms to compare. This is done by means of an atomic similarity threshold, which needs to be surpassed, and a final gradient optimization to refine the alignment. More details on this algorithm can be found in the work of Constans et al [128].

Figure 2.2 shows, as an illustrative example, the alignment of steroids aldosterone and methyl-fluorocortisol obtained using the maximal similarity algorithm described above. As can be observed in the figure, the maximal similarity search leads to a superposition of the steroid ring system, three hexagonal and one pentagonal rings, which constitute the common backbone atom skeleton of the two steroids.

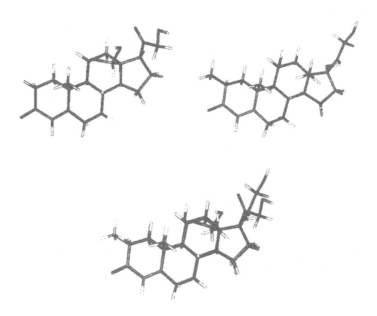

Figure 2.2. Alignment of aldosterone and methyl-fluorocortisol using the maximal similarity rule.

2.11.2.1
Superpositions involving heavy atoms

Due to the fact that products between densities make MQSM, the contribution at each point is strongly sensitive to the existence of atomic nuclei, since high-density peaks exist at those positions. Taking this into account, the superposition algorithm might find as the optimal alignment the exact superposition of two heavy atoms, regardless of the remaining structure.

As an example of this fact, the superposition of two industrial pesticides, namely ofurace and orylazin, is shown in Figure 2.3. The heaviest atom in ofurace is a chlorine atom, whereas in orylazin is a sulfur. The maximal similarity is obtained when these two atoms are exactly superimposed, aligning the rest of atoms in such a way that the contribution to the total MQSM is increased as much as possible. The common substructures, namely the benzene ring, are not superimposed.

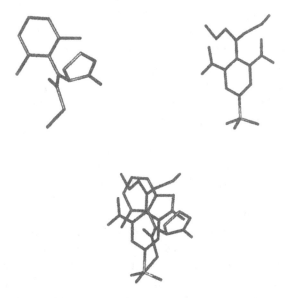

Figure 2.3. Alignment of ofurace and orylazin according to the maximal similarity rule.

2.11.3
Common skeleton recognition: the topo-geometrical superposition algorithm

As it has been commented, the molecular alignment problem can be faced from different perspectives. Together with the maximal similarity rule, another possible method developed in this laboratory consists of searching the maximal common substructure between two molecules. This approach uses only topological and geometrical molecular features, leading to the so-called *topo-geometrical superposition algorithm* (TGSA). In this approach, MQSM are not used during the alignment process, and only interatomic distances and atom types are considered.

TGSA is based on the comparison between interatomic distances and atomic numbers, providing a way to align molecular pairs, which ensures the superposition of the common skeleton. This procedure is connected to the *maximum isomorphic subgraph problem* studied in pattern recognition and statistics [129].

The algorithm is structured as follows. The first step defines all the atomic bonds, storing the interatomic distances in a database. The total number of bonds is the number of atomic dyads considered in the superposition process. Hydrogen atoms are not included throughout the calculations.

Once all dyads are defined, each dyad of a molecule is compared to those of the second molecule. Two dyads are considered to be similar if their distances are equal within a given threshold (5% in bond length). This step allows discarding a considerable amount of dyads that do not belong to the common backbone atoms. Starting from the surviving dyads, atomic *triads* are constructed by adding a bonded atom to a dyad. Triads are, in geometrical terms, triangles, whose vertices correspond to atoms and two or three of their sides correspond to chemical bonds. All triads of a molecule obtained with this procedure are compared with all triads of the other one, using again the comparison of the distances with the same threshold. If two triads are found to be similar, the three atoms of each molecule are considered to constitute a common skeleton fragment.

Reached this point, two situations can occur: no pair of triads are found to be similar, or several pairs are selected. In the first case, the threshold value is increased and the preceding steps are repeated. The second situation requires the definition of a criterion of triad selection. The adopted criterion consists of choosing the pair of triads that superposes the major number of equal atoms from the whole molecule, that is, the alignment representing the largest common substructure. If several triads superpose the same number of equal atoms, the one that minimizes the sum of distances between the superposed atoms will be selected.

In order to illustrate the kind of superpositions obtained with TGSA, the same example as before, i.e. the molecular pair (ofurace, orylazin), will be analyzed. As Figure 2.4 shows, the common benzene ring is exactly superposed. Note the difference with Figure 2.3, produced by the fact that the density peaks of heavy atoms do not participate in the alignment.

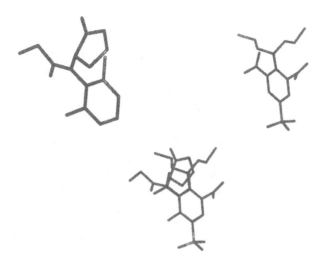

Figure 2.4. Alignment of ofurace and orylazin using TGSA.

2.11.4
Other molecular alignment methods

The first methods of molecular superposition have been issued in the 70's, with the works of Gavuzzo et al [130] and McLachlan [131]. The work of McLachlan was continued by Gerber and Muller [132], and later, by Redington [133]. All these studies were based on the minimization of an error measure defined as the sum of the distances between the atomic positions of the two compared structures. On the other hand, alternative methods were proposed taking the values of different physico-chemical properties as the superposition background. These properties usually are fields that simulate steric or electrostatic effects, either calculated in three-dimensional grids [134-136] or approximated with analytical functions [137,138]. The use of similarity indices as a source of the alignment has been proposed by other authors, by means of optimizing the involved indices with Monte Carlo techniques [139] as well as with gradient methods [140]. Finally, the superposition of dissimilar molecules has also been discussed [141].

3 Application of Quantum Similarity to QSAR

One of the most progressing subjects in present-day chemistry is the establishment of quantitative relationships between biological or pharmacological properties and molecular structure. This topic has become a solid subject matter, usually known as *quantitative structure-activity relationships* (QSAR). Since Hansch and Fujita [142] performed the pioneering studies on QSAR, the advances in this matter have not ceased. The predictive capabilities of the earliest models were substantially improved when 3D structural descriptors were introduced, providing a powerful alternative to the use of *extra-thermodynamical* parameters in QSAR studies [143]. In addition, the definition of different quantitative similarity measures between two molecules proved a great aid in order to a source of 3D QSAR parameters acting as molecular descriptors.

Many QSAR methods have been recently proposed [144-148], most of them employing different molecular descriptors and dealing with different statistical tools [149-155]. In this chapter, a scheme of the application of quantum similarity descriptors to examine structure-molecular properties relationships of interest is exposed. MQSM and QSAR are deeply interconnected by quantum-mechanical theoretical structure, as it has been sketched in the previous chapter and will be later discussed. Only a general view of the application of Quantum Similarity to QSAR is given in the following sections, leaving the different specific applications to be analyzed in detail in next chapters.

3.1
Theoretical connection between QS and QSAR

Due to the particular form of the quantum similarity measures, and their quantum mechanical nature, a theoretical connection between the first-order density function and a physical observable can be established. This association can be extended to the similarity vectors and to several plausible transformations of them.

3.1.1
Beyond the expectation value

A formal connection between molecular properties and structural descriptors generated by the MQSM can be set. It can be based upon one of the most fundamental principles in Quantum Mechanics, which relates the system density function to any physical observable through an appropriate Hermitian operator set up. Known the state density function, all observable property values of the system can be formally extracted from it, by means of the expectation value of the associated Hermitian operator, which acts over the corresponding density function, within a formal structure borrowed from theoretical statistics:

$$y = \langle \Omega \rangle = \int \Omega(\mathbf{r}) \rho(\mathbf{r}) d\mathbf{r} = \langle \Omega | \rho \rangle \qquad (3.1)$$

Being y the observable studied, and Ω an unknown quantum operator, considered being non-differential. The equations to construct the predictive models reported in this work constitute a discretization of the expectation value rule as set in equation (3.1), where a transformation of the corresponding similarity vector play the role of the density function:

$$y_I = \langle \Omega \rangle \approx \boldsymbol{\beta}^T f(\mathbf{z}_I) \qquad (3.2)$$

First derived by Carbó et al. [156], such a connection acts as a fundamental equation in all QSAR issuing from MQSM computations. For this reason, one can refer to QSAR built up in this way as *quantum QSAR* or *QQSAR*.

3.2
Construction of the predictive model

As in all QSAR techniques, the practical application of an equation as (3.2) is performed starting with a set of compounds of known activity, which is used as a first step to build a predictive model. This model is then applied to predict the property for a series of molecules of unknown activity. The following procedure is coincident with the classical QSAR algorithms.

The usual form to elaborate a predictive model utilizes a set of objects whose property values are known. These objects serve as a calibration or *training set* for the model. Thus, a training of the model, which relates the descriptors to the measured properties, is induced. This leads to simple mathematical equations able to associate properties and descriptors as best as possible. If the training set is a representative sample of the set, then it is assumed that the introduction of new elements possessing an unknown property value, the *test set*, will not affect its stability, and that predictions of enough confidence can be obtained.

In this work, it has been preferred to describe a methodology that emphasizes the premise of simplicity. Nowadays, there exists a general tendency to construct sophisticated QSAR models where complex variable selection processes and training set correlation are employed. In the procedures described next, a special effort has been provided in order to keep maximal simplicity as far as possible. This is done by imposing linear transformations in the relationships between the descriptors and the properties, assuming the derived cost in terms of predictive capacity loss.

3.2.1
Multilinear regression

For the sake of simplicity, and owing to the underlying quantum theoretical procedures outlined in section 3.1.1, the relationship between the molecular descriptors and the physico-chemical or biological property is imposed to be linear. Thus, the relationship is presented here in the form of a kth-order multilinear regression (MLR) [157], using the transformed similarity measures as parameters:

$$\mathbf{y} = \beta_0 \mathbf{1} + \mathbf{X}_{(k)} \beta_{(k)} + \varepsilon_{(k)} \tag{3.3}$$

where $\mathbf{y}$ is a n-dimensional vector containing the properties or activities of the studied molecular family, $\mathbf{1}$ is a unity vector, that is a column vector composed of unit elements, $\mathbf{X}(k)$ denotes the reduced $(n \times k)$ configuration matrix and $\varepsilon_{(k)}$ is a normally distributed error term having zero expectation value and dispersion matrix $\mathbf{I}\sigma^2$. The $\{\beta\}$ estimators are computed using an ordinary least-squares technique. It is appropriate to simplify the predictive model, so it is intended to construct regressions involving the minimal number of descriptor parameters (k). This objective justifies the use of variable selection techniques. In some cases, only one parameter ($k=1$) will be used, which leads to the simplest linear regression.

Several different descriptors can be used as parameters; however, all of them will be derived here from the Quantum Similarity framework. Examples of these descriptors are: principal coordinates of the system, common and transformed self-similarities, or electron-electron repulsion energies. Details on the specific form and the way of derivation of these descriptors will be discussed in the following chapters.

3.3
Possible alternatives to the multilinear regression

The linear association between physical observables and molecular descriptors is not unique. There are other two multivariate analysis techniques widely used in QSAR analysis: *partial-least squares* (PLS) regression and *neural network algorithms*. A brief overview of both techniques is exposed here.

3.3.1
Partial least squares (PLS) regression

A widely used alternative to multilinear regression is the so-called partial least squares (PLS) projection to latent structures analysis method [149]. It is particularly adequate when the number of variables exceeds largely the number of objects, as occurs in the 3D QSAR methods built on function grids. PLS is a robust linear method that assumes that there is a small number of variables, which really matter: the latent variables (LVs), and that all the structural descriptors can be viewed as combinations of these LVs plus some error. The connection with the property is established by means of a multilinear regression using the LVs as parameters. PLS is able to deal with collinear data, that is, it can use a great amount of *correlated* variables. Such a technique has been successfully used in MQSM calculations on molecular families with a common moiety [158].

3.3.2
Neural Network algorithms

Artificial neural networks constitute a non-linear approach usually employed in QSAR analysis [159]. These algorithms try to simulate the neurological processing of the human brain. A neural network is made up of different layers, divided into three types: an input layer, (various) hidden layers and an output layer. A non-linear transference function is defined from the input data, weighted by coefficients that are adjusted according to the resultant output. The training of the network is supervised through the minimization of an error function defined from the difference between the desired and the obtained values. The process is repeated by back-propagation until self-consistency is attained. A possible problem of this methodology is some high risk of overparametrization. Moreover, in MQSM approach such as the one followed in this work, the fundamental linear relationships, coming from quantum mechanical considerations, make neural network gear out of context.

3.4
Parameters to assess the goodness-of-fit

Once the regression is constructed, a quantitative indication of the accuracy of the property estimation is needed. This is traditionally done by means of two statistical parameters: the multiple determination coefficient r^2 and the standard deviation σ_N.

3.4.1
The multiple determination coefficient r^2

The multiple determination coefficient is defined as:

$$r^2 = 1 - \frac{\sum\limits_{i=1}^{n}(y_i - \hat{y}_i)^2}{\sum\limits_{i=1}^{n} y_i^2 - \frac{1}{n}\left(\sum\limits_{i=1}^{n} y_i\right)^2} = 1 - \frac{SS_E}{S_{yy}} \qquad (3.4)$$

Where SS_E is the residual sum of squares, and S_{yy} the total sum of squares. This coefficient was compelled to vary between 0 and 1; the closer to 1, the more similar are the adjusted values to the experimental ones. The limiting case is obtained when all the residuals are zero, and the model adjusts perfectly to the data. The definition, presented in equation (3.4), is simply the multidimensional extension of the *Pearson coefficient*, defined for linear regressions of a single variable [157]. It must be noted, however, that a high r^2 coefficient does not imply a good predictive model: the simple addition of a parameter to the regression increases the value of r^2, even when the new descriptor does not contribute to the predictivity of the model. To determine the predictive capacity of the model, other measures are used, which will be later defined.

3.4.2
The standard deviation coefficient σ_N

The standard deviation of the data is defined as:

$$\sigma_N = \sqrt{\frac{1}{n}\sum\limits_{i=1}^{n}(y_i - \hat{y}_i)^2} \qquad (3.5)$$

This coefficient provides a non-normalized measure of the dispersion of the data estimation. As it can be seen, it is the scaled root form of the r^2 numerator. The squared sum of residuals is a widely used factor in many statistical parameters, but its usual form in statistics involves the squared differences between the data and their mean [157]. The former is also known as *standard deviation of errors of calculation* (SDEC).

3.5
Robustness of the model

Together with the quantification of the goodness-of-fit, it is very important to obtain a measure of the model predictive capacity and stability. Since the elaboration of the model is based on a series of objects with a known property, it can occur that one or more elements of the set may possess structural particularities that made them different from the rest. These elements will determine the exact form of the model, and their presence is essential.

3.5.1
Cross-validation by leave-one-out

The most usual way to explore the stability of a predictive model is through the analysis of the influence of each one of the individual objects that configure the final model. To do this, one object of the set is extracted, and the model is recalculated using as training set the remaining $n-1$ objects. The property is then predicted for the removed element. This process is repeated for all the objects of the set, obtaining a prediction for every one. This procedure is known as *cross-validation* by *leave-one-out* [160]. Obviously, this process can be extended to a larger set of removed elements, yielding the leave-two-out, leave-three-out... or general leave-many-out procedures. These are even more decisive for robustness assessment. Nevertheless, they are rarely reported within current QSAR results, and they have not been used in this work.

3.5.2
The prediction coefficient q^2

A factor measuring the dispersion of these predictions is computed from the cross-validated values for each object. This factor is the *predictive residual sum of squares*, denoted by the acronym PRESS:

$$\text{PRESS} = \sum_{i=1}^{n} \left(y_i - \hat{y}_{(i)} \right)^2 \tag{3.6}$$

The "hat" of the variable y, as is the usual statistical notation, indicates that it is a predicted value of the studied property, and the parenthesis in the subindex indicates that the prediction has been made by leave-one-out cross-validation. The PRESS function has not to be confused with the squared sum of residual employed in the definition of r^2: in the latter, the difference was computed between the experimental property and the *adjusted* value, not the predicted one.

The PRESS factor is used to define the coefficient of prediction q^2:

$$q^2 = \frac{SD - PRESS}{SD} \qquad (3.7)$$

SD is the squared deviation of the observed value from their mean:

$$SD = \sum_{i=1}^{n} (y_i - y)^2 \qquad (3.8)$$

It must be noted that, in contrast to r^2, the coefficient of prediction can take on values less than zero. Let us suppose that one assigns to each individual a predicted property value equal to the arithmetic mean of the experimental data. The residuals of this assignment are measured with the SD factor, defined in equation 3.8. If the achieved errors within the model built, given by the PRESS factor, are larger than the former ones, the expression SD–PRESS becomes negative, and so the q^2 function. This is the interpretation of the possible negative values of q^2. A value q^2>0.5 is commonly accepted as a satisfactory result.

In contrast to the r^2 coefficient, which augments when more parameters are added to the regression and is non-negative defined, the factor q^2 exhibits a curve with a maximum at a certain number of parameters, which monotonically decreases afterwards. This fact can be easily explained: increasing the number of parameters of the model always improves the data adjustment, but this fact makes not necessarily the model more predictive. An extreme situation is attained when the model has as many parameters as objects; in this case the adjustment is perfect –it is a compatible determined system–, but it does not possess any predictive capacity. In fact, the descriptors of the model could be random numbers, which would be subsequently perfectly adjusted to the data. This fact confers a great importance to the q^2 coefficient.

The usual way to represent the cross-validation results is by plotting the experimental properties versus the cross-validated ones. The X and Y-axes range the same values, and a diagonal straight line crosses the plot. The closer are the points to the diagonal line, the better is the description. On the other hand, if the points present a great dispersion, the model becomes a poor one.

3.5.3.
Influence on the regression results

There are certain aspects that can determine the model results and which can be avoided by a previous data analysis. Some of these features are briefly described in this section.

3.5.3.1
Data distribution

It is very important to get an adequate scaling of the experimental data. Most of the biological activities are measured in terms of concentrations or curative/lethal doses, and they can vary within a wide range. Using a logarithmic transformation, which is often employed to reduce the huge differences between data values, usually scales such property types. A smooth and regular property distribution is necessary to get reliable results. Otherwise, if the property is made up of isolated islands, *clusters*, of data points, some fitting problems can arise. Thus, apparently satisfactory results can be obtained simply whether the multilinear regression passes through these islands, even if the intercluster differences are not described. These apparently good results can easily mask a poor estimation of the activity.

3.5.3.2
Outliers

The results of the multilinear regression are highly influenced by the presence of *strange* points. The q^2 factor is quite sensitive to bad data descriptions, and only one point with a large cross-validation residual leads to a low q^2 value. These points are usually called *outliers* of the system. In the Quantum Similarity framework, while remembering the definition of MQSM, those molecules with heavy atoms can produce a bad alignment of the data. The misalignment may possibly produce a poor description of the corresponding point. Thus, it is necessary to examine the superposition patternss generated by the training set, especially if there are one or more points with a different structure. In this sense, it must be emphasized that the homogeneity of the studied sets is essential.

If an outlier is detected, the QSAR researcher has to decide what to do with it. The easiest solution consists of removing the implied molecule from the set. It is important, however, to have a convincing argument to do this. A point should only be removed from the set if its poor description can be unequivocally associated to structural reasons. If the QSAR model cannot predict satisfactorily the property of a molecule with no structural particularities, then it is the methodological background what needs to be revised.

3.6
Study of the chance correlations

In any predictive model, excellent adjustments and even satisfactory predictions can be achieved without the existence of a real relationship between the molecular structure description and the property of the studied set. A method capable to distinguish between a real structure-activity correlation and a chance description is then needed in every case.

3.6.1
The randomization test

Although the nature of Quantum QSAR is such that a causal relationship always may be supposed to exist, via the fundamental equation, in order to evidence the existence of fortuitous correlations, the *randomization test* is adopted in this book [161]. This test consists of building a property vector y_{RND} whose components are the components of the actual property vector, but randomly permuted in their positions, in this way:

$$\mathbf{y} = \left(y_1, y_2, ..., y_n\right)^T \xrightarrow{RND} \mathbf{y}_{RND} = \left(y_8, y_5, ..., y_2\right)^T \tag{3.9}$$

This new activity vector is used as if it was really an experimental one, and a QSAR model is computed in the usual way. This process is repeated several times –currently, from 70 to 100 times–, in order to test the capacity of the model to extract actual structure-activity relationships. If statistically significant models are achieved when using the randomly ordered properties, the model must be considered suspect of random correlation. It probably correlates any data set, due to an excess of either parameters or degrees of freedom. Nevertheless, a detailed analysis is necessary in those cases where good models are found out for wrong properties, since the routine of generation of permuted vectors might have created an altered vector very similar to the real one.

The usual form to represent the results of a randomization test is by plotting a two-dimensional figure using the correlation and prediction coefficients as axes. Two different geometrical symbols are used to differentiate those points that represent the real property and those that correspond to the permuted ones. Figure 3.1 shows two examples of this kind of representation. The first one corresponds to a system which has succeeded the randomization test, whereas the second one corresponds to a model able to correlate unreal data.

3.7
Comparison between the QSAR models based on MQSM and other 2D and 3D QSAR methods

Various methodologies are being currently applied to the analysis of molecular properties. These methodologies utilize different molecular descriptors and statistical techniques, but they can be classified into two general kinds of problem approach: *two-dimensional* (2D) methods, which use physico-chemical properties or molecular topological features as descriptors; and *three-dimensional* (3D) methods built on grids, which compute relevant molecular fields or properties in a set of points surrounding the examined molecules.

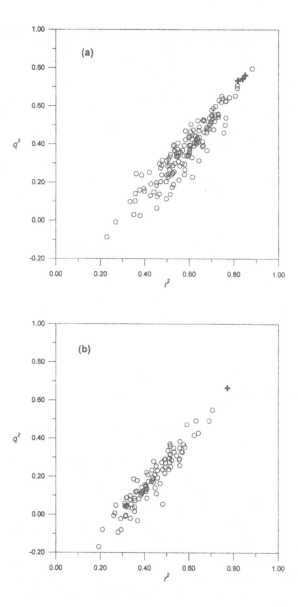

Fig. 3.1. Examples of randomization tests. Points corresponding to randomly ordered properties have been marked with a circle, and those corresponding to the actual property have been marked with a cross. **(a)** The model is able to adjust and predict satisfactorily unreal data. **(b)** The model gives good results only when actual data is analyzed.

3.7.1
Comparison with 2D methods

The initial QSAR approaches have used molecular physico-chemical properties as descriptors, related linearly to the studied property [143]. Several descriptors, namely the *octanol/water partition coefficient* log P or *Hammett's* σ, provided good results in a wide range of situations, especially when congeneric or highly homogeneous systems were treated. However, the large number of descriptors available, and the introduction of indicator variables to explain outlying deviations in particular cases, has transformed the choice of parameters within this classical QSAR methodology into a quite obscure process. Besides this, most physico-chemical properties have several different parameters to represent them, leading to a considerable confusion in their selection.

Other different 2D techniques have been proposed in the QSAR environment, based on molecular topological features [162]. Several topological indices were noted to correlate satisfactorily some molecular properties, and since then, new indices have been added to a great collection of topological descriptors [163,164]. Both approaches present an unsolved difficulty: the election of the appropriate descriptor for each particular case. In the majority of cases, this is made in an arbitrary way. Besides this, many physico-chemical properties are interrelated, and therefore redundant. On the other hand, quantum similarity provides a consistent, homogeneous and unbiased set of parameters, independent of the molecular properties available.

3.7.2
Comparison with 3D methods built on grids

In some cases, the two-dimensional parameters are not enough to represent the determinant variables for activity. Historically, three-dimensional effects were introduced for the first time to describe a molecular set difficult to correlate by traditional 2D methods, a set of steroids that binds to the corticosteroid-binding globulin. The approach employed to solve this problem, the Comparative Molecular Field Analysis (CoMFA) [144], is widely used at this moment. CoMFA is based on the calculation of relevant molecular energy fields (steric, electrostatic, hydrophobic) at every point of a three-dimensional rectilinear grid surrounding the molecules. The matrices derived in this way are then dealt with statistical techniques, mainly PLS, yielding to predictive models. Some slight modifications and improvements to the original Cramer work have been proposed since then, see for example [165-168].

In most cases, the QSAR approach based on quantum similarity measures is able to obtain comparable results to other highly predictive QSAR models, even when they use more sophisticated non-linear neural network techniques. However, the main advantages of the model proposed here are mainly related to the molecular descriptor choice. Quantum similarity measures, constructed as indicated in previous sections, are better than those derived from the construction of stereoelectronic maps, done by the evaluation of relevant molecular energy fields in a 3D rectilinear grid surrounding the compound. This is so because MQSM overcome obvious dependencies on grid parameters (size, spacing and location), on the probe-atom chosen and on the relative orientation of the molecule within the chosen grid. Another problem never discussed in such procedures is related to the grid's shape. The examination of the different possibilities for the grid form, namely rectilinear, cylindrical, spherical, and so on, for a better adaptation to the molecular structure has not been done yet in detail. Moreover, the usual overlap and Coulomb MQSM are more robust descriptors, respectively, than the usual measurement of the Lennard-Jones 6-12 and Coulomb potentials at the grid points. This is so because these exhibit singularities at atomic positions, and this is corrected by setting to zero the potentials at these points. The usual procedures also avoid the truncation at arbitrarily fixed cut-off values near the atomic positions due to the extremely large contributions of the potentials there. Consistency in the results of grid-based QSAR approaches is a problem discussed by various authors [169-171], and different improvements to overcome these drawbacks have been proposed [165-168].

3.8
Limitations of the models based on MQSM

Once the general procedure for correlating quantum similarity data to molecular properties has been exposed, it is necessary to discuss the limitations of the approach.

3.8.1
Homogeneity of the sets

General molecular similarity techniques, and quantum similarity approach in particular, are based on the extraction of the information contained in the resemblance measures between different compounds. Quantum Similarity uses density function superposition as the basis for the building of the comparative measures. The usual QSAR procedure consists of the analysis of the effect of a series of ligands in the protein structure. If the effect on a given molecule is well known, it can be observed that the molecular structure may suffer slight alterations and the consequences of these changes are easily evaluated. The structural modifications in a given set of molecules to be studied are commonly changes in a small number of substituents or rings. This usually yields to studies on compounds of a same family. For an optimal density superposition, it is interesting that the molecular set presents itself as homogenous as possible. If the compounds possess a common skeleton, this tends to be optimally superposed, and the quantum similarity information is then mainly referred to the comparison of the dissimilar fragments arranged among the molecules of the set.

3.8.2
The problem of the bioactive conformation

One of the most powerful characteristics of the QSAR framework is the possibility to quantify the effect of a molecular set in a biochemical reaction involving a protein, without any direct information of the substrate. In 3D QSAR methods, however, the molecular geometry is an essential feature for the construction of the descriptors. The atomic coordinates of the molecules should be chosen in such a way that the real conformation of the ligand when interacts with the substrate is reproduced. This feature is called the *bioactive conformation*, and it must be noted that this conformation has not necessarily to match the minimum energy one. Data from experimental results, involving a large number of compounds, are not always available, and sometimes the minimum energy geometry has to be unavoidably used. In such cases, the reliability of the derived results must be examined with caution, since the chosen conformation determines the constructed models.

3.8.3
Determination of molecular alignment

In the quantum similarity framework, as well as in many of the 3D QSAR approaches built on grids, the molecular alignment is a sensitive variable to be fixed. The main difficulty is that the researcher arbitrarily imposes the molecular alignment, since there no exists a unique definition of *optimal* superposition. However, as it has been previously discussed, the alignment rule usually employed in the quantum similarity approach is clear and simple: two molecules will be superposed in such a way that the value of the chosen similarity integral measure is maximal. This procedure tends to align the common skeleton of the ligands. Nevertheless, when heavy atoms are present in the structure of the molecules, the density peak produced by the exact match of two heavy atoms might be significant enough to determine the superposition of the rest of the compound. In any case, the algorithm used allows for the possibility to match three pair of atoms of the analyzed compounds, and therefore the common fragments can be aligned per force. A different perspective to the problem is provided by the topo-geometrical algorithm, which searches the maximal common substructure of the molecules under comparison, regardless of the atomic composition.

4 Full molecular quantum similarity matrices as QSAR descriptors

In this chapter, a scheme of the application of molecular quantum similarity matrices to describe a molecular property of interest is exposed. Quantum similarity matrices need to be conveniently transformed when employed as descriptor source in QSAR procedures. In order to describe the usual transformations, dimensionality reduction and variable selection techniques will be discussed. Combination of different quantum similarity matrices, constituting the Tuned QSAR model, is also discussed. Since the only relevant test for the procedure protocol is its application on real cases, quantum similarity matrices will be used to study three different molecular sets in order to provide the reader with reliable quantitative equations for activity prediction.

4.1
Pretreatment for quantum similarity matrices

Quantum similarity matrices, as defined in chapter 2, cannot be straightforwardly used as molecular descriptors in a QSAR environment, due to the high number of descriptors involved in a similarity matrix, equal to the number of molecules analyzed. To adapt them to the usual QSAR procedures, several previous treatments are necessary.

4.1.1
Dimensionality Reduction

Research in Multivariate Analysis seems to coincide into affirm that *multidimensional scaling* (MDS) techniques are the most appropriate for the treatment of similarity and dissimilarity –that is, *proximity*– data (see for example, [172]). Working with ($n \times n$) similarity matrices is excessive for the present QSAR purpose, and some transformation is necessary to reduce the dimensionality of the data while loosing the minimal original information possible.

The different models for the reduction of dimensions of similarity matrices search a common final objective: they pretend to map the initial proximities $\{p_{ij}\}$ into distances $\{d_{ij}\}$ of a multidimensional space. The coordinates in this new multidimensional space are collected into the coordinate matrix $\mathbf{X}$. The mapping between proximities and distances is specified by means of a representation function f:

$$f : p_{ij} \rightarrow d_{ij}(\mathbf{X}),$$ (4.1)

Or, more generally:

$$f : g(p_{ij}) \rightarrow d_{ij}(\mathbf{X}),$$ (4.2)

Where g (p_{ij}) symbolizes any transformation of the original proximities. In the present case, the proximities p_{ij} correspond to the elements Z_{ij} of the similarity matrix $\mathbf{Z}$, and $g(p_{ij})$ could be the matrix elements of any MQSI.

The relationship induced by the function f cannot be exactly fulfilled. In the practical case, $\mathbf{X}$ configuration is analyzed in such a way that the transformation f is satisfied as best as possible, *i.e.* that the interpoint distances match as close as possible the original proximities. The condition "as best as possible" has a mathematical translation through the aggregation of mapping errors for each individual. The way to define this measure, called *loss of information function*, or more commonly, *loss function*, yields to the different dimensionality reduction techniques. As it can be noted, these methods have a clear geometrical origin [173].

For the sake of simplicity, in the following discussion p_{ij} will denote the quantitative proximity measure between two objects, encompassing the original matrix elements Z_{ij} or the possible transformations generated by the different MQSI. Thus, it will not be necessary to specify if the proximity is a C-class or D-class similarity index.

The dimension reduction techniques attempt to represent the proximities as interpoint distances in a multidimensional space, assumed to have dimension m. Point coordinates of the configuration are referred to a set of m orthogonal axes that intersect at a point-origin O. Each point i in this space is univocally described by a m-tuple as:

$$\mathbf{x}_i = (x_{i1}, x_{i2}, ..., x_{im}),$$ (4.3)

where $\mathbf{x}_{ia}$ is the projection of object i on axis a. This m-tuple is known as the i-th *principal axis* or *principal coordinate* (PC) of the system. The origin of coordinates O is taken as the zero vector $\mathbf{0}=(0,0,...,0)$.

Because of the application space image f is considered to be Euclidean, the distance between two points corresponds to the length of the segment which joins them. The extension of the definition of distance between points in other non-Euclidean metrics will not be analyzed in this work.

Mathematically, the distance between two points i and j belonging to an Euclidean space can be easily computed by means of the usual expression:

$$d_{ij}(\mathbf{X}) = \left[\sum_{a=1}^{m} (x_{ia} - x_{ja})^2 \right]^{1/2} \tag{4.4}$$

Thus, $d_{ij}(\mathbf{X})$ is equal to the sum of the squared differences $x_{ia}-x_{ja}$. As it has been specified in all the given equations, the interpoint distance depends on the configuration $\mathbf{X} = (x_1, x_2, \ldots x_n)^T$. The last expression can be rewritten in a norm-like way:

$$d_{ij}^2(\mathbf{X}) = (\mathbf{x}_i - \mathbf{x}_j)(\mathbf{x}_i - \mathbf{x}_j)^T = \| \mathbf{x}_i - \mathbf{x}_j \|^2 \tag{4.5}$$

4.1.1.1
Classical Scaling

Classical scaling was one of the first multivariate analysis tools for similarity data treatment. Its origins have to be found in psychometrics, where it provided a valuable aid to understand the judgments of people on the similarity of a series of objects. Torgerson [174] proposed the first MDS method and coined the term. This work was an evolution of Richardson's approach [175], who utilized the theorems of Eckart and Young [176] and Young and Householder [177]. Nowadays, this technique is applied in all those fields working with proximity matrices, such as psychology, psychiatry, sociology, marketing, political science, biology, physics and chemistry.

The central idea of this approach considers the proximities as distances, and then tries to obtain coordinates able to explain them. In order to develop the classical scaling formalism, the problem can be faced inversely, that is, let us suppose that the coordinate matrix $\mathbf{X}$, of dimensions $(n \times m)$, is known for a set of n points in Euclidean space. The way to construct a squared distance matrix starting with this matrix is simple:

$$\mathbf{D}^{(2)} = \mathbf{c}\mathbf{1}^T + \mathbf{1}\mathbf{c}^T - 2\mathbf{X}\mathbf{X}^T = \mathbf{c}\mathbf{1}^T + \mathbf{1}\mathbf{c}^T - 2\mathbf{B}, \tag{4.6}$$

Being $\mathbf{B} = \mathbf{X}\mathbf{X}^T$, also $\mathbf{1}$ is a unity vector: an n-dimensional vector of ones, and $\mathbf{c}$ is a vector having as components the diagonal elements of $\mathbf{B}$. Multiplying both sides by the factor $-1/2$, and by the centering matrix $\mathbf{J} = \mathbf{I} - n^{-1} \mathbf{1}\mathbf{1}^T$, it is obtained:

$$\begin{aligned}
-\tfrac{1}{2}\mathbf{J}\mathbf{D}^{(2)}\mathbf{J} &= -\tfrac{1}{2}\mathbf{J}(\mathbf{c}\mathbf{1}^T + \mathbf{1}\mathbf{c}^T - 2\mathbf{X}\mathbf{X}^T)\mathbf{J} \\
&= -\tfrac{1}{2}\mathbf{J}\mathbf{c}\mathbf{1}^T\mathbf{J} - \tfrac{1}{2}\mathbf{J}\mathbf{1}\mathbf{c}^T\mathbf{J} + \tfrac{1}{2}\mathbf{J}(2\mathbf{B})\mathbf{J} \\
&= -\tfrac{1}{2}\mathbf{J}\mathbf{c}\mathbf{0}^T - \tfrac{1}{2}\mathbf{0}\mathbf{c}^T\mathbf{J} + \mathbf{J}\mathbf{B}\mathbf{J} = \mathbf{B}
\end{aligned} \tag{4.7}$$

The first two terms vanish because centering a vector of ones yields a vector of zeros ($\mathbf{1}^T \mathbf{J} = \mathbf{0}$). The centering on $\mathbf{B}$ has not any influence, because $\mathbf{B}$ is already centered.

The solution which classical scaling provides, pretends to relate interpoint distances in a multidimensional space to primitive proximities. In this sense, the solutions of the method are invariant under global data translations, therefore it is necessary to adopt some criterion for selecting the origin of coordinates. The double centering allows the components of each column of matrix $\mathbf{X}$ to sum zero:

$$\sum_{i=1}^{n} x_{ia} = 0, \qquad a = 1,...,m \tag{4.8}$$

Subsequently, the origin of coordinates of the configuration coincides with the center of gravity (centroid) of the considered n-dimensional points.

The matrix $\mathbf{B}$ can be calculated by taking the quantum similarity matrix $\mathbf{Z}$ (or any MQSI) as a distance matrix $\mathbf{D}$, and conveniently transforming it as indicated in equation (4.7). Recovering the coordinates $\mathbf{X}$ from the matrix $\mathbf{B}$ can be done through the spectral decomposition:

$$\mathbf{B} = \mathbf{X}\mathbf{X}^T = \mathbf{V}\,\mathbf{\Lambda}\,\mathbf{V}^T, \tag{4.9}$$

where $\mathbf{V}$ is the orthogonal matrix containing the eigenvector and $\mathbf{\Lambda}$ is a diagonal matrix which has the eigenvalues of $\mathbf{B}$ as no-null elements. Coordinate matrix $\mathbf{X}$ is simply:

$$\mathbf{X} = \mathbf{V}\mathbf{\Lambda}^{1/2} \tag{4.10}$$

The eigenvectors are arranged in a descending order according to their associated eigenvalue

4.1.1.2
The loss function in classical scaling: the Strain function

From a variational point of view, classical scaling can be considered to minimize a loss of information function called *Strain*, defined as:

$$L(\mathbf{X}) = \left\| \mathbf{X}\mathbf{X}^T - \left(-\tfrac{1}{2} \mathbf{J}\mathbf{Z}\mathbf{J} \right) \right\|^2 \tag{4.11}$$

An interesting property of classical scaling consists of that the dimensions are nested. This means, for instance, that the first two dimensions of the three-dimensional solution are exactly the same that the unique two dimensions of the two-dimensional solution. This fact implies that once the similarity matrix to be treated is defined, classical scaling will only need to be computed once, independently of the dimensionality of the predictive model constructed. This does not occur with other MDS methods.

In relation to the use of MQSM as a source of QSAR parameters, classical scaling is a very powerful method, due to the fact that it generates a big set of variables capable to provide several different subsets for correlation.

4.1.1.3
The problem of the diagonal elements

It has been stated that classical scaling considers the proximities as distances, and therefore, it neglects the diagonal elements, which are taken to be zero. In the particular case of the quantum similarity matrices, the diagonal elements, which coincide with the self-similarities, are not null, and even they are different one to another. Hence, the neglect of self-similarities in classical scaling may lead to a considerable loss of information.

There are two different ways to face this problem. On one hand, it can be assumed the loss of information derived from this approach, or, on the other hand, the original similarity matrices are transformed into MQSI matrices, which possess diagonal elements which are identical one to another, and therefore, without containing any relevant information.

4.1.1.4
Dissimilarities and Euclidean distances. Gower's transformation

On the other hand, let $\Delta = \{d_{ij}\}$ be an Euclidean distance matrix. By definition, it satisfies the following expressions:

$$
\begin{aligned}
d_{ij} &\geq 0 \quad \forall i,j \\
d_{ii} &= 0 \quad \forall i \\
d_{ij} + d_{ik} &\geq d_{jk} \quad \forall i,j,k
\end{aligned}
\tag{4.12}
$$

It can be proved [178,179] that when a matrix of this kind is transformed by means of classical scaling, always exists a solution where the interpoint distances in the multidimensional space match *exactly* the original distances. In this case, the matrix $\mathbf{B} = \mathbf{X}\mathbf{X}^{\mathrm{T}}$ is positive semidefinite, that is, all its eigenvalues are positive or zero. Furthermore, the same theorem demonstrates that the rank of the solution is at most $n-1$. It must be emphasized that only Euclidean distance matrices satisfy this property, but it is not true for all the dissimilarity matrices, even when they fulfil the first two conditions of equation (4.12).

Furthermore, starting from a similarity matrix $\mathbf{S}$, where $0 \leq S_{ij} \leq 1$ and $S_{ii} = 1$; Gower and Legendre [180] demonstrated that the transformation:

$$d_{ij} = |1 - s_{ij}| \tag{4.13}$$

produces an Euclidean distance matrix, which subsequently satisfies all the previous properties.

There exists, therefore, two ways to deal with the proximity matrices between quantum objects. The first one consists of working with the Gower transformation of the Carbó indices, and then transforming the distances as described in equation (4.7) to be dealt with classical scaling. Joining both transformations yields:

$$-\tfrac{1}{2} d_{ij}^2 = -\tfrac{1}{2}\left[|1 - s_{ij}| \right]^2 = -\tfrac{1}{2}\left(1 - s_{ij}\right) = \tfrac{1}{2}\left(s_{ij} - 1\right) \tag{4.14}$$

As can be observed, merging both equations produces a linear transformation of the Carbó indices. Thus, using together these two changes, or directly the Carbó indices, will not produce any influence on the projection final result. With respect of the application of MQSM in the QSAR environment, the difference between both treatments will lead to differences in the concrete numeric form of the predictive model, the regression estimators; but not in the quality of the models, quantified by the statistical coefficients of correlation and prediction, previously defined in chapter 6.

4.1.2
Variable selection

Once the dimensionality of the original quantum similarity matrix is reduced, it is necessary to select the principal axes (PCs) that will constitute the QSAR model descriptors. There exist different criteria to choose the variables.

4.1.2.1
Selection independent of the external variables

When a transformation such as classical scaling is carried out, the resulting configuration has as many components as the cardinality of the original matrix. This large number of potential descriptors needs to be filtered, since only a small subset of PCs will be used to build the QSAR models. The most obvious selection is choosing those k axes accounting for the maximal variance, measured by each eigenvalue, since this selection provides the k-dimensional subspace that best fits the original data.

$$x_1 \succ x_2 \succ ... \succ x_k \text{ following } \lambda_1 \succ \lambda_2 \succ ... \succ \lambda_k \tag{4.15}$$

This selection is very usual, but it presents some problems. The main problem consists of that in this scheme it is implicitly assumed that the analyzed property best correlates with those PCs describing in the best way the most external differences between the members of the set. As can be seen, the external data, that is, y variables, do not intervene at all in this variable selection technique.

4.1.2.2
Selection dependent on the external variables: the most predictive variables method

In the realistic application of MQSM to QSAR, an effective variable selection technique is needed. The technique presented in this subsection, and applied in the next examples, is called the *most predictive variables method* (MPVM) [181,182], and consists of arranging the PCs according to a measure of their predictivity, which involves the molecular property examined. A predictivity measure can be defined simply by evaluating the projection of each PC on the data:

$$\chi^2(\mathbf{x}_i, \mathbf{y}) = \frac{(\mathbf{y}^T \mathbf{x}_i)^2}{\sum_j (y_j - y)^2 \lambda_j} \qquad (4.16)$$

$\mathbf{y}$ is the property vector, $\mathbf{x}_i$ is the i-th PC and λ_i the eigenvalue associated to this PC. The denominator of this expression is a constant that has the same value for each PC. On the other hand, the numerator gives a measure of the projection of the i-th PC on the property data, which becomes the criterion for ordering the descriptors. The PCs will be arranged according to the value of this coefficient:

$$\chi^2(\mathbf{x}_{i_1}, \mathbf{y}) > \ldots > \chi^2(\mathbf{x}_{i_p}, \mathbf{y}) \qquad (4.17)$$

A k-th order determination coefficient can be defined as:

$$X_{(k)}^2 = \sum_{\alpha=1}^{k} \chi^2(\mathbf{x}_{i_\alpha}, \mathbf{y}) \qquad (4.18)$$

The k variables chosen will be those which make the coefficient $X^2_{(k)}$ maximal.

In order to avoid obtaining relationships involving very low variance PCs, which is a signal of background noise parametrization, a variance threshold has been adopted. Thus, all those PCs accounting for eigenvalues lower than an arbitrary percentile value will be rejected. This threshold is usually chosen to be 1%. In any case, model validations are still necessary to verify that real structure-activity relationships are found out.

4.2
The MQSM-QSAR protocol

Up to now, all the elements for the construction of a QSAR predictive model based on the full Quantum Similarity matrix are given. The procedure follows the following steps:

- Calculation of the molecular geometry, either by experimental (X-ray diffraction) or computational (*ab initio*, semiempirical, molecular mechanics) methods.
- Construction of the molecular densities. To avoid expensive computational calculations, the ASA approach is used.
- Computation of the MQSM, conveniently optimized and collecting them in a matrix.
- Reduction of dimensions and selection of variables
- Construction of the multilinear regression and assessment of the quality of the regression. Validation of the models.

All the previous indications can be summarized in an illustrative flowchart, shown in figure 4.1.

4.3
Combination of quantum similarity matrices: the tuned QSAR model

So far it has only been considered the use of single quantum similarity matrices as a source for QSAR parameters. In this section, it will be introduced the possibility of combining two or more quantum similarity matrices at a time. It will be also detailed the constraints that must fulfil the mixing coefficients in order to keep the quantum mechanical structure of the MQSM. General instructions for computing the optimal coefficients will be briefly discussed.

4.3.1
Mixture of matrices and coefficient constraints

The particular structure of MQSM allows the possibility of combining them to lead new measures. The mixed MQSM will utilize different matrices derived from several different positive definite operators. The ultimate mathematical and quantum mechanical justifications for these mixtures are based in the convex set formalism.

The combination of matrices will be constructed simply by adding weighted elements to a sum:

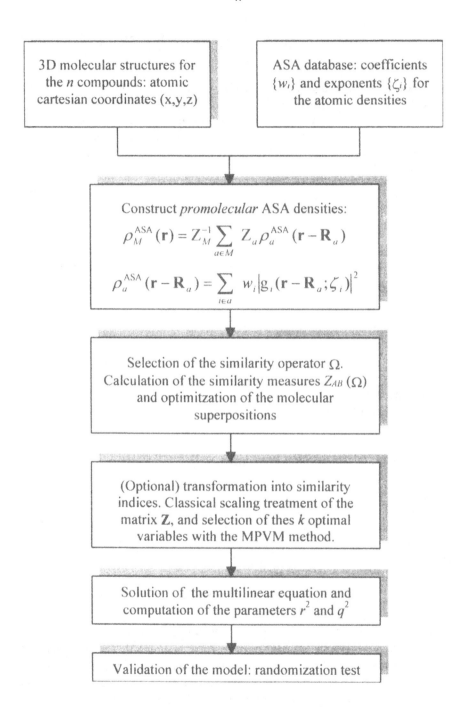

Fig. 4.1. Flowchart of the MQSM-QSAR protocol.

$$\mathbf{Z}^{tuned} = c_1\mathbf{Z}(\Omega_1) + c_2\mathbf{Z}(\Omega_2) + ... + c_m\mathbf{Z}(\Omega_m) = \sum_{\varepsilon=1}^{m} c_\alpha\mathbf{Z}(\Omega_\alpha) \qquad (4.19)$$

where each Ω_α corresponds to a different quantum similarity operator. In this work, combinations of overlap and Coulomb operators will be only used.

In order to avoid scale factors when combining different quantum similarity matrices, a previous scaling is adopted. The scaling is simply made by dividing each matrix element by the sum of all the elements:

$$Z_{AB}(\Omega) = \frac{Z_{AB}(\Omega)}{\sum_{I,J} Z_{IJ}(\Omega)} \qquad (4.20)$$

This scaling makes easier the interpretation of the matrix weights.

In order to guarantee that the new structure maintains the MQSM properties, the coefficients $\{c_\alpha\}$ must fulfill the following constraints:

$$c_\alpha \geq 0 \quad \forall\alpha$$
$$\sum_{\alpha=1}^{m} c_\alpha = 1 \qquad (4.21)$$

These constraints are known as the *convex conditions*. They ensure that the new tuned quantum similarity matrices keep all the elements positive.

4.3.2
Optimization of the convex coefficients

A clear criterion for choosing the weights for each matrix is needed. As this approach will be employed in the QSAR analyses, it will be precisely the robustness measures of the predictive models that will determine their exact form. Thus, the coefficients of the similarity matrices combined will be those that *maximize the coefficient of predictivity q^2*:

$$\{c_\alpha\} \text{ such that } q^2 = \text{maximal} \qquad (4.22)$$

Finding the exact numeric expression for these weights is not an easy task. To do this, a practical algorithm was developed by Amat *et al.* [183]. The computational algorithm encompasses two stages: first, a global search is carried out by means of a Monte Carlo procedure [184]. A random sequence of weights is generated, and the predictive model is constructed following the steps given in the previous section. The q^2 value obtained is stored. Starting with this sequence, a Fibonacci succession [185] is applied to refine the search. This cycle is repeated until a satisfactory QSAR model is achieved. Another possible degree of freedom of the model consists of the number of combined matrices. As only a combination of μ matrices from a total of ν possible ones is wanted, a systematic search among the total number of existing possibilities of combination, given by a simple combinatorial factor, is performed. The implementation of this combinatorial search is carried out by means of the so-called *nested summation algorithm* [186,187].

4.4
Examples of QSAR analyses from quantum similarity matrices

In order to illustrate the predictive capacity of the QSAR models derived from the analysis of quantum similarity matrices, three molecular sets have been studied, covering three different application fields for QSAR techniques: medicinal chemistry, toxicology and protein engineering. These case examples have to be added to other papers on the subject, in which quantum similarity matrices were successfully correlated to molecular properties [43-45,47,51,52].

4.4.1
Activity of indole derivatives

The most important application field of QSAR analyses is the study of the activity of pharmacologically active compounds. Molecules that intervene in biochemical processes constitute a huge chemical set with a broad interest ranging from chemistry to medicine passing through biology and pharmacy. The particular difficulty in modeling the drug-receptor interactions makes interesting an alternative approach to the problem: the theoretical relationships between descriptors associated with the molecular structure and their biological activity. The major part of QSAR studies reported, since the creation of QSAR techniques, tries to explain the trends in activity for molecular sets of pharmacological interest.

Thus, the first application example presented in this study is related to this subject. A set of 23 indole derivatives and their capacity to displace [^{3}H] flunitrazepam from binding to bovine brain membrane will be analyzed using Quantum Similarity tools. The indole family is found to bind to the benzodiazepine receptor. Benzodiazepines are the most frequently prescribed drugs for the pharmacotherapy of anxiety, of *status epilepticus* and convulsive, emotional disorders, so they possess a notable pharmacological interest. Further, the design of agonists or antagonists for this receptor might be of a great importance for medicinal chemists. The indole activity is measured with pK_i, K_i being the concentration necessary to produce 50% inhibition in an assay with bovine brain membrane. A previous study on this same data set by Hadjipavlou-Litina and Hansch [188], using classical descriptors, already existed. The results of these authors will be compared with the ones obtained with MQSM techniques.

Table 4.1 gathers the general structure of the studied indole derivatives, together with the substituents and the activity data. All the substitutions are similar, involving hydrogen and chlorine atoms, as well as OH, NO_2 and OCH_3 groups. The common skeleton for this family is considerably large, and therefore this set seems adequate for being studied with Quantum Similarity techniques.

For the MQSM calculations, theoretically computed geometries have been used. Thus, geometry optimization using the semiempirical AM1 Hamiltonian [189] was first carried out. Density functions were then constructed using promolecular ASA approach. Coulomb operator has been used for MQSM, transformed afterwards into Carbó indices. In order to construct the prediction regressions, the protocol shown in Fig. 4.1 was used, including classical scaling and MPVM analyses. The derived results are given in Table 4.2, where the optimal model has been marked in bold face.

Table 4.1. Indole derivatives. Identification number, substituents and activity.

No	R, R_1, R_2, R_3	pK_i	No	R, R_1, R_2, R_3	pK_i
1	H, H, H, H	6.93	13	H, Cl, H, H	7.17
2	Cl, H, H, H	6.21	14	H, H, H, Cl	5.59
3	NO_2, H, H, H	6.93	15	H, OH, H, H	6.37
4	H, OCH_3, H, H	6.78	16	Cl, OH, H, H	6.82
5	Cl, OCH_3, H, H	6.68	17	NO_2, OH, H, H	7.92
6	NO_2, OCH_3, H, H	7.27	18	H, H, OH, H	6.09
7	H, H, OCH_3, H	6.54	19	Cl, OH, H, H	6.24
8	Cl, H, OCH_3, H	6.79	20	NO_2, OH, H, H	7.19
9	NO_2, H, OCH_3, H	7.42	21	H, OH, OH, H	6.46
10	H, OCH_3, OCH_3, H	7.03	22	Cl, OH, OH, H	6.75
11	Cl, OCH_3, OCH_3, H	7.52	23	NO_2, OH, OH, H	7.32
12	NO_2, OCH_3, OCH_3, H	7.96			

Table 4.2. Statistical coefficients for the predictive QSAR models using MQSM. The optimal model has been marked in bold face.

Number of PCs	Selected PCs	r^2	q^2	σ_n
1	3	0.349	0.246	0.454
2	3, 2	0.648	0.556	0.334
3	**3, 2, 1**	**0.761**	**0.631**	**0.275**
4	3, 2, 1, 5	0.786	0.628	0.260

As it can be noted from Table 4.2, the best model is found when using 3 PCs, leading to satisfactory adjustments. The number of descriptors is small enough to consider the model as not very complex, and the derived results are of an appreciable quality. In addition, the variable selection chooses the first three PCs, and consequently, the explained variance of the model coincides with the best three-dimensional similarity subspace. The equation for the optimal model can be written as:

$$y = -4.03x_3 - 3.52x_2 - 1.61x_1 + 6.87 \qquad (4.23)$$

In order to assess the robustness of the model, the leave-one-out cross-validation procedure was performed. Thus, each molecule of the set was extracted and a model formed by the remaining 22 indoles predicted its activity. The predicted pK_i values for the optimal model are given in Table 4.3.

Table 4.3. Experimental and cross-validated activities for the 23 indole derivatives.

ID	Observed pK_i	Predicted pK_i	ID	Observed pK_i	Predicted pK_i
1	6.93	6.33	13	7.17	7.00
2	6.21	6.51	14	5.59	5.00
3	6.93	7.28	15	6.37	6.77
4	6.78	6.87	16	6.82	6.74
5	6.68	6.92	17	7.92	7.38
6	7.27	7.65	18	6.09	6.50
7	6.54	6.52	19	6.24	6.56
8	6.79	6.52	20	7.19	7.25
9	7.42	7.25	21	6.46	6.81
10	7.03	6.97	22	6.75	6.79
11	7.52	6.88	23	7.32	7.51
12	7.96	7.60			

As can be easily observed, there are not present large deviations between the experimental and predicted activities. The highest error is accounted for molecule **14**: 10.5%. On the other hand, this compound possesses the lowest activity, 5.59, a little bit far from the rest of compounds of the set. Figure 4.2 shows the experimental activities against the predictions by cross-validation. The molecule associated to the less active point is indole **14**. Its particular location, placed far from the point cloud, can overestimate the capacity of the model. This fact was pointed out to be an influent factor in regression results (see section 3.5.3). The poor prediction for compound **14** can be explained in this manner: when this molecule is removed from the set in the leave-one-out procedure, the influence of it in the regression disappears and the regression coefficients may change considerably. The new coefficients, describing the point cloud, are not good representatives of the distant molecule. Indole activity data is therefore an illustrative example of the influence of data distribution in the regression results, and must be taken into account when discussing the results. A treatment for this problem will be given below.

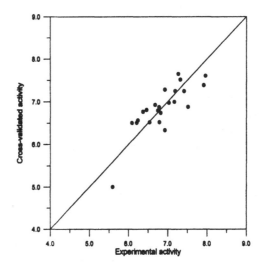

Figure 4.2. Experimental versus cross-validated activities for the 23 indole derivatives.

3D QSAR models, due to their complexity and the sophistication of the employed chemometric tools, can be a dangerous source of chance correlations. In order to assess that the obtained relationships are not due to chance correlations and overparametrizations, a validation by means of the randomization test was carried out. 100 new activity vectors were generated from the permutation in the position of the actual vector components. These altered activities were then used as a data source for QSAR models under the optimal conditions given by 3 PCs. The results for the randomized models can be compared with the real starting one only by representing in a plot the statistical coefficients r^2 and q^2. This is depicted in Figure 4.3. The statistics for the modified activity vectors are clearly lower than the real QSAR model, and for the major part a result of $q^2<0$ is obtained. This ensures that a real structure-activity relationship has been found out.

The present results cannot be straightforwardly compared with those obtained by Hadjipavlou-Litina and Hansch [188]. These authors used Hammett's σ as the principal descriptor for correlation. Hammett's σ constant quantifies the substituent effect on a series of chemicals with a common structure, as in the indole case. To improve the description, two indicator variables were added to the regression by Hadjipavlou-Litina and Hansch, which took discrete values according to the presence-absence of a given type of substituent. However, these three variables were not descriptive enough to correlate satisfactorily this system, and three molecules were removed by the mentioned authors from the set: **2**, **13** and **14**. The best result for the 20-indole model yielded r^2=0.810.

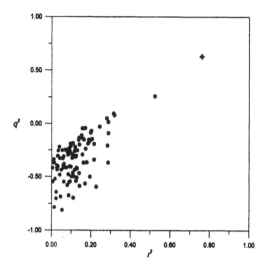

Figure 4.3. Randomization test associated to the previous QSAR model. Circles represent the randomly ordered activities, and the cross corresponds to the real activities.

It has already been stated that one of the points can have a high influence in the QSAR model results. Thus, following the reasoning discussed in the theoretical section, the molecule of lower activity, located far from the point cloud, can artificially increase the goodness-of-fit. To evaluate this effect, molecule **14** was extracted from the set and the regressions were recalculated, while keeping the same conditions as before: Coulomb operator, Carbó indices, most predictive variables method. The results achieved are compiled in Table 4.4.

Table 4.4. Statistical coefficients for the predictive QSAR models using MQSM. Molecule 14 has been removed. The optimal model has been marked in bold face.

Number of PCs	Selected PCs	r^2	q^2	σ_n
1	2	0.341	0.197	0.408
2	2, 3	0.523	0.366	0.347
3	**2, 3, 4**	**0.698**	**0.535**	**0.276**
4	2, 3, 4, 1	0.723	0.517	0.265

Again the optimal model is obtained when using 3 PCs. As expected, the quality of the regressions has sensibly decreased; confirming that indole **14** produced an artificial correlation in the previous model. The equation for the new 3 PCs model is:

$$y = -3.29x_2 + 2.87x_3 + 3.04x_4 + 6.93. \tag{4.24}$$

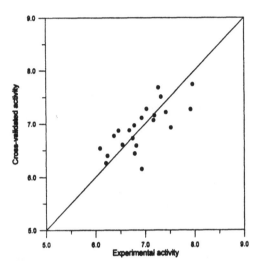

Figure 4.4. Experimental versus cross-validated activities for the indole derivative set. Molecule **14** has been removed.

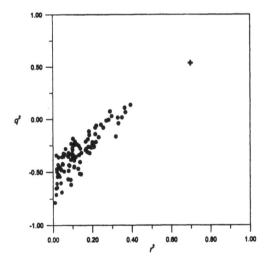

Figure 4.5. Randomization test associated to the previous QSAR model with molecule **14** removed. Circles represent the randomly ordered activities, and the cross correspond to the real activities

Exactly as before, figure 4.4 shows a plot contrasting experimental and cross-validated pK_i. Now the point dispersion is higher than before, but the general trends discerning between high and low active compounds are still present in the plot. Finally, figure 4.5 gathers the randomization test for the optimal QSAR model. Again the model succeeds in the validation tests, and a clear separation between statistics corresponding to real and randomized responses is observed. As a conclusion, although the new results are less valuable, they still possess chemical significance, because the model is only able to correlate real data.

4.4.2
Aquatic toxicity of substituted benzenes

Another subject where currently QSAR techniques are being successfully applied is the study of the toxicity of certain compounds to animal and vegetal species [190,191]. These studies are of a great scientific and social interest. Structure-toxicity relationships have become useful tools to get a better understanding of the molecular mechanisms of many chemical and biochemical processes, as well as to propose simple mathematical expressions for the prediction of toxic properties. Another useful derivation of the QSAR studies in toxicity is the possibility to constitute an alternative, or at least, a complement, of toxicological assays with animals.

The common QSAR studies in toxicology utilize 2D descriptors such as the octanol-water partition coefficient (log P) or other parameters, which simulate the different molecular interactions. It is commonly accepted that the action of the pollutant compounds on the animal species is made through a disruption in the functioning of cell membranes. Thus, the toxic capacity of such a molecule is deeply connected to the tendency to accumulate in the cellular membranes. A simple model to describe the cell membranes can be done by using an apolar medium, namely octanol. Then, it is not surprising that satisfactory QSAR models are built when using the octanol/water partition coefficient (log P) [192]. In fact, the inverse relationship between the toxic action and log P is commonly known as *baseline toxicity*. To improve the description of these models, other 2D descriptors are usually employed, such as the HOMO and LUMO energies, or parameters that describe the ionic and covalent contributions of the hydrogen bonding.

Quantum Similarity is proposed here as a possible tool to find out hidden structure-activity relationships in a toxicity study. Recently, it was proved that there exists a clear linear relationship between log P and the molecular quantum self-similarity measures (MQS-SM) for highly homogeneous series [193]. This connection will be more deeply analyzed elsewhere ahead of this book. Therefore, if log P correlates satisfactorily with pollutant toxicity, then MQS-SM, and, in extension, MQSM, should also adequately describe these agents too. The main advantage of this methodology is that do not require a previous selection of the descriptors used, a process that is sometimes made rather arbitrarily in classical QSAR.

Table 4.5. Compounds and aquatic toxicity (LC50) for the 92 narcotic pollutants.

Compound	LC50	Compound	LC50
Chlorobenzene	-3.77	4-*n*-butylphenol	-4.47
1,2-dichlorobenzene	-4.40	4-*tert*-butylphenol	-4.46
1,3-dichlorobenzene	-4.28	2-*tert*-butyl-4-methylphenol	-4.90
1,4-dichlorobenzene	-4.56	4-*n*-pentylphenol	-5.12
1,2,3-trichlorobenzene	-4.89	4-*tert*-pentylphenol	-4.81
1,2,4-trichlorobenzene	-4.83	2-allylphenol	-3.96
1,3,5-trichlorobenzene	-4.74	2-phenylphenol	-4.76
1,2,3,4-tetrachlorobenzene	-5.35	1-naphtol	-4.50
1,2,3,5-tetrachlorobenzene	-5.43	4-chlorophenol	-4.18
1,2,4,5-tetrachlorobenzene	-5.85	4-chloro-3-methylphenol	-4.33
3-chlorotoluene	-3.84	4-chloro-3,5-dimethylphenol	-4.66
4-chlorotoluene	-4.33	3-methoxyphenol	-3.22
2,4-dichlorotoluene	-4.54	4-methoxyphenol	-3.05
2,4,5-trichlorotoluene	-5.06	4-phenoxyphenol	-4.58
3,4-trichlorotoluene	-4.60	Quinoline	-3.63
Pentachlorotoluene	-6.15	Aniline	-2.91
Benzene	-3.09	2-methylaniline	-3.12
Toluene	-3.13	3-methylaniline	-3.47
2-xylene	-3.48	4-methylaniline	-3.72
3-xylene	-3.45	*N,N*-dimethylaniline	-3.33
4-xylene	-3.48	2-ethylaniline	-3.21
Nitrobenzene	-2.97	3-ethylaniline	-3.65
2-nitrotoluene	-3.59	4-ethylaniline	-3.52
3-nitrotoluene	-3.65	4-butylaniline	-4.16
4-nitrotoluene	-3.67	2,6-diisopropylaniline	-4.06
2,3-dimethylnitrobenzene	-4.39	2-chloroaniline	-4.31
3,4-dimethylnitrobenzene	-4.21	3-chloroaniline	-3.98
2-chloronitrobenzene	-3.72	4-chloroaniline	-3.67
3-chloronitrobenzene	-4.01	2,4-dichloroaniline	-4.41
4-chloronitrobenzene	-4.42	2,5-dichloroaniline	-4.99
2,3-dichloronitrobenzene	-4.66	3,4-dichloroaniline	-4.39
2,4-dichloronitrobenzene	-4.46	3,5-dichloroaniline	-4.62
2,5-dichloronitrobenzene	-4.59	2,3,4-trichloroaniline	-5.15
3,5-dichloronitrobenzene	-4.58	2,3,6-trichloroaniline	-4.73
2-chloro-6-nitrotoluene	-4.52	2,4,5-trichloroaniline	-4.92
4-chloro-2-nitrotoluene	-4.44	4-bromoaniline	-3.56
Phenol	-3.45	α,α,α,4-tetrafluoro-3-methylaniline	-3.77
2-methylphenol	-3.77	α,α,α,4-tetrafluoro-2-methylaniline	-3.78
3-methylphenol	-3.48	Pentafluoroaniline	-3.69
4-methylphenol	-3.74	3-benzyloxyaniline	-4.34
2,4-dimethylphenol	-3.86	4-hexyoxyaniline	-4.78
2,6-dimethylphenol	-3.75	2-nitroaniline	-4.15
3,4-dimethylphenol	3.92	3-nitroaniline	-3.24
2,3,6-trimethylphenol	-4.21	4-nitroaniline	-3.23
4-ethylphenol	-4.07	2-chloro-4-nitroaniline	-3.93
4-propylphenol	-4.09	4-ethoxy-2-nitroaniline	-3.85

The molecular set studied as a toxicity case example is a series of 92 aromatic compounds, which possess acute toxicity to the fish *Poecilia reticulata*. The toxicity was measured as the lethal concentration necessary to reduce 50% the initial population: the so-called LC50 factor. Experimental data reported by Verhaar *et al.* [194] will be used. This large molecular set is made up of benzene, toluene, xylene, phenol and aniline derivatives. The different molecules and their corresponding activity are shown in Table 4.5. These compounds have been extracted from a larger set studied by Urrestarazu *et al.* [195] employing a 2D QSAR techniques.

The geometries of all the studied compounds have been optimized at the semiempirical AM1 computational level [189] with the program AMPAC 6.0 [196]. The particular structure of these molecules, namely a substituted hexagonal ring, made unnecessary expensive *ab initio* calculations. Coulomb MQSM and Carbó indices were used. The dimensionality of the quantum similarity matrix was reduced with classical scaling, and the optimal PCs were selected using the most predictive variables method with a 1% variance threshold.

4.4.2.1
Quantitative structure-toxicity relationships

This molecular set was studied using the MQSM-QSAR protocol described previously. Dealing the Carbó index matrix with the exposed methodology, the results collected in table 4.6 were achieved. As can be seen, valuable models are obtained using few PCs. The q^2 coefficient slightly augments with the increase of parameters, and some criterion must be adopted to choose which constitutes the optimal model. The way to do it is by using a plot of the predictive coefficient evolution against the number of descriptors. This kind of representation is called *scree plot* [197], and it is shown in figure 4.6. This plot brings forward information on the description generated by the different models, and points out a possible criterion for truncation: the optimal number of parameters corresponds to an *elbow* in the plot. This elbow shows the point when the addition of new parameters does not contribute significantly to the predictivity of the model. In this particular case, the 3 PCs model can be considered as the best one, yielding a value $q^2 = 0.716$. It must be noted that the optimal variables are the first three ones. That is, those that represent the three-dimensional subspace that best fit the original Carbó indices.

The optimal predictive model follows the expression:

$$\overset{.}{y} = -2.248x_1 + 1.443x_2 + 3.281x_3 - 4.138 \qquad (4.25)$$

Table 4.6 QSAR model results for the 92 narcotic pollutants.

# PCs	Selected PCs	Explained Variation	r^2	q^2	σ_N
1	1	26.95 %	0.389	0.358	0.508
2	1, 3	36.25 %	0.675	0.651	0.371
3	**1, 3, 2**	**47.24 %**	**0.740**	**0.716**	**0.331**
4	1, 3, 2, 5	54.55 %	0.761	0.733	0.318
5	1, 3, 2, 5, 4	63.05 %	0.770	0.737	0.319
6	1, 3, 2, 5, 4, 14	64.58 %	0.793	0.758	0.296

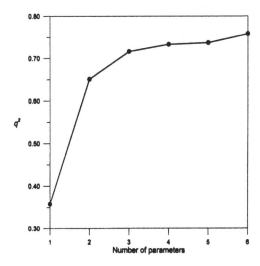

Fig. 4.6. Scree plot for the predictive coefficient: evolution of q^2 with the number of parameters for the 92 pollutants.

Some molecules are not very well described by the model, possessing residuals higher than 0.70: pentachlorobenzene, 2-*tert*-butyl-4-methylphenol, 4-*n*-pentylphenol, 1-naphtol, 4-methoxyphenol and 2-ethylaniline. The variable selection algorithm chooses some axis apparently not relevant in terms of the explained variance. Indeed, the difference between the explained variance of the 6-variables model which includes the 14th PC and the one resulting from employing the first 6 variables is only 3.68%, clearly negligible. In other words, the presence of apparently higher PCs, such as the 14th, is not a problem, since the variance explained by this axis is similar to many of the lowest PCs.

Figure 4.7 shows the observed toxicities versus the cross-validated ones. As can be seen, the point dispersion is reasonably small, taking into account the number of molecules treated.

The randomization test, shown in figure 4.8, clearly ensures the existence of a linear relationship between the toxicity of the aromatic compounds and the descriptors derived from the MQSM. As can be observed, the permuted responses yield poor predictive models, all having $q^2<0.2$. On the other hand, the correctly ordered toxicities yield satisfactory statistical parameters, and therefore it is located isolated in the plot. From this representation can be concluded that the obtained results are not produced by chance correlations or by an overparametrization of the multilinear regression. Furthermore, these results indicate another interesting fact: MQSM, accounting for shape features of the molecules, are also able to describe in a satisfactory way a property related to molecular hydrophobicity. The possibility for Quantum Similarity to find out valuable QSAR models out of its "natural" environment, that is, drug-receptor interactions where steric effects are the main responsible for differences in the ligand activity, opens the way to a more general application of MQSM and enhances the fundamental, theoretical connection between Quantum Similarity and QSAR, exposed in the previous chapter.

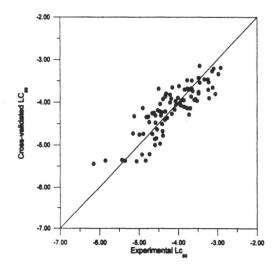

Fig. 4.7. Experimental versus cross-validated toxicities for the optimal 3 PCs model.

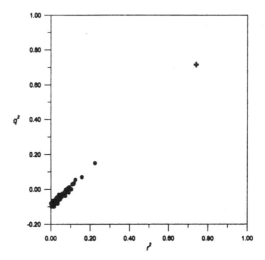

Fig. 4.8. Randomization test for the 92 aromatic compounds using the 3 PCs optimal model. Permuted responses have been marked with circles, and the correctly ordered toxicities with a cross.

4.4.2.2
Qualitative structure-toxicity relationships

Together with the QSAR analysis, it may be interesting to carry out a qualitative discrimination between the high and low toxic compounds. This kind of analysis is more reliable than the quantitative one, and it is useful whether only a binary (low/high), ternary (low/medium/high) or higher classification level is searched. For this purpose, in the particular case of the 92 aromatic pollutants, the toxic action to *Poecilia reticulata* has been divided into two classes: high and low toxicity.

The distribution of the toxicity in this molecular set does not contain intuitive groups. This means that the property levels cannot be fixed in an intuitive way, and therefore a criterion of discretization needs to be adopted. Size homogeneity in classes was imposed to be this criterion. Thus, in order to get groups of comparable size, the threshold between the high and low active levels has been fixed to 4.1 units. The continuity on the toxicity values can lead to misclassifications of those compounds which possess a toxic action close to the threshold chosen.

The way to analyze qualitatively the Quantum Similarity data is simply by projecting the point-molecules on the optimal classical scaling subspace. The PCs taken as axes are those that the MPVM method selects as the optimal ones. The two- or three-dimensional graphical representation brings forward information on the Quantum Similarity criteria for closeness between molecules; and the location of the narcotic pollutants could then be connected to the discrete value of the toxic action. Due to the satisfactory quantitative results previously reported, a separation between both toxicity classes is expected.

First, a two-dimensional plot of the molecular set is created, given in Fig. 4.9. In this picture, three groups are distinguished, which have been marked with dashed ellipses. In a first approximation, the molecules are separated according to its toxicity. Thus, the cluster located at the northwest of the plot contains the major part of low toxic agents. On the other hand, the remaining groups contain the majority of highly toxic compounds. As can be observed, there are some misclassifications in this picture: 16 compounds, of a total of 92, are located in a region that does not correspond to its toxicity. Some of these misclassifications can be explained by the neglected information when the two-dimensional solution is employed. In this sense, adding a third axis to the plot may overcome some of these deficiencies. Figure 4.10 shows the three-dimensional classical scaling configuration using the three most predictive PCs as axes. This plot removes one misinterpretation of the previous one: five of the bad classified molecules in the rightmost region possess a high projection on the third axis (2nd PC), whereas the actual highly toxic agents present a very low projection, even negative. Thus, the three-dimensional representation helps to consider in a more realistic way the grouping generated by MQSM. It can be concluded that the major part of the compounds, almost 88%, is correctly classified within this simple scheme. In addition, several incorrectly discriminated molecules are narcotic pollutants that possess a toxic action close to the interclass threshold. It has been commented that the class boundaries have been chosen arbitrarily, so these molecules could have been included in the opposite group if an alternative criterion were adopted.

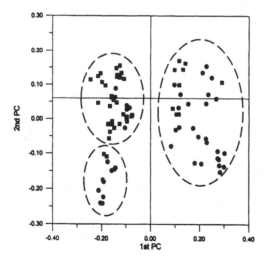

Fig. 4.9. 2D classical scaling solution associated to the substituted benzenes using the two optimal PCs as axes. Low toxic action has been marked with a square, and high toxicity with a circle.

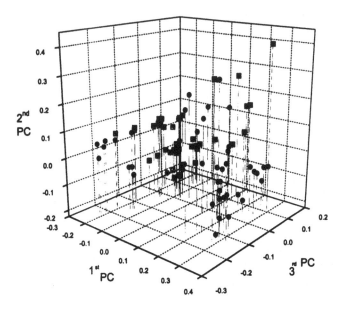

Fig. 4.10. 3D classical scaling solution associated to the substituted benzenes using the three optimal PCs as axes. Molecules with low toxic action have been marked with a square, and those with high toxicity with a circle.

4.4.3
Single-point mutations in the subtilisin enzyme

One of the most promising subjects in protein engineering is the design of enzymes with desired properties adequate for determined chemical reactions [198]. Site-directed mutagenesis experiments provide a suitable tool to evaluate the importance of a particular aminoacid for the activity, specificity or stability of the protein [199]. The interpretation of the results of these experiments is often made in a qualitative and empirical way. As an alternative to this, there have recently appeared the quantitative structure-function (QSFR) and structure-stability (QSSR) relationships, which are currently being tested on several different systems and with different approaches [200-204]. These methodologies are, in fact, the natural analogues of QSAR techniques for this class of problems.

The method used in this section is closer to that used in the previous ones. Here, the studied molecules are the different aminoacids substituted at a given position of a protein. The aminoacids are homogeneous compounds, and therefore, they can be dealt with Quantum Similarity techniques.

First of all, it is necessary to clearly distinguish between the processes that intervene in QSAR and those that intervene in the QSFR-QSSR techniques. There exist certain analogies between both concepts, but in some aspects they are describing opposite processes. In QSAR, the reaction of a given ligand on some substrate is well known, and slight modifications in the ligands are carried out, evaluating their effect on the activity. These modifications are usually changes in the substituents at a few positions, considered being relevant for activity. Alternatively, in QSFR-QSSR, starting with the known ligand-substrate reaction, some punctual modifications in the protein are introduced, and the effect of these changes on its function/stability is then studied. These modifications are single-point changes of any aminoacid located in the active site of the enzyme.

As it has been commented, site-directed mutagenesis searches an improvement of the protein properties by modifying certain aminoacids that are relevant for the biochemical reaction [205]. In the present section, controlled mutations on subtilisin enzyme and their effect on the function of the whole molecule are analyzed. Subtilisin is one of the most studied enzymes, and a great amount of site-directed mutagenesis experiments have been carried out on this protein [206-209]. One of the purposes for preparation of mutants is the improvement of their stability and catalytic efficiency, necessary for an industrial application. The work of Stauffer and Etson [210] implied the residual Met222 as a primary center for the oxidative inactivation of subtilisin. Estell et al [211] prepared 19 mutants of subtilisin with the cassette mutagenesis technique created by Wells et al [212]. The objective was to find the optimal replacement in the active site to design an enzyme resistant to chemical oxidation. The property analyzed was the inhibition relative activity of the oxidative center of the different mutants, assayed in the substrate N-succinyl-L-Ala-L-Ala-L-Pro-L-Phe-p-nitroanilide 0.3mM. Activity data were taken from the mentioned work of Estell et al.

Table 4.7 shows the activity values for the wild-type and 19 mutations of subtilisin. The property was scaled using the logarithmic form in order to avoid problematic differences in the values.

The amino acid set is usually contained in the database of several commercial chemistry software products. In the current example the aminoacid structures have been obtained from the SPARTAN program database [213]. The molecular density functions have been built using ASA approach. Overlap MQSM, transformed into Carbó indices have been used. Carbó index matrix has been treated with classical scaling and the optimal PCs have been selected using the most predictive variables method.

To quantify the influence of the deliberate changes in the protein structure, the quantum similarity matrix has been processed with the protocol described in this chapter. The different models are shown in Table 4.8. As it can be noted, the models are quite satisfactory, and the number of descriptors is reasonably small. The 4 PCs model has been considered the optimal one, yielding values of $r^2=0.764$ and $q^2=0.620$.

Table 4.7. Amino acids and subtilisin activity.

Amino acid	Acronym	Activity (K)	pK
Alanine	Ala	53.0	-1.724
Arginine	Arg	0.5	0.301
Asparagine	Asn	15.0	-1.176
Aspartate	Asp	4.1	-0.613
Cysteine	Cys	138.0	-2.140
Glutamine	Gln	7.2	-0.858
Glutamic Acid	Glu	3.6	-0.556
Glycine	Gly	30.0	-1.477
Histidine	His	4.0	-0.602
Isoleucine	Ile	2.2	-0.342
Leucine	Leu	12.0	-1.079
Lysine	Lys	0.3	0.523
Methyonine	Met	100.0	-2.000
Phenylalanine	Phe	4.9	-0.690
Proline	Pro	13.0	-1.114
Serine	Ser	35.0	-1.544
Threonine	Thr	28.0	-1.447
Tryptophan	Trp	4.8	-0.681
Tyrosine	Tyr	4.0	-0.602
Valine	Val	9.3	-0.968

Table 4.8. QSFR models for subtilisin mutations. The optimal model has been marked in bold face.

# PCs	Selected PCs	r^2	q^2	σ_N
1	1	0.326	0.214	0.546
2	1, 6	0.521	-0.068	0.460
3	1, 6, 7	0.678	0.384	0.377
4	**1, 6, 7, 4**	**0.764**	**0.620**	**0.324**
5	1, 6, 7, 4, 8	0.802	0.488	0.296

The multilinear regression for the optimal model is:

$$y = -1.424x_1 + 2.017x_6 - 1.889x_7 - 1.074x_4 - 0.939 \qquad (4.26)$$

Figure 4.11 shows the graphical representation of the experimental activities versus the cross-validated ones. The point dispersion is reasonably small, although in this case there are three points a little bit far away from the rest. The poor description of one of the extreme points, corresponding to the lowest activity mutation (Lys), can have a decisive influence in the value of the prediction coefficient q^2. The cross-validation residual attributed to this point is large: 0.843.

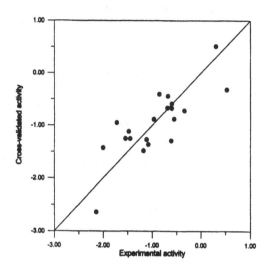

Fig. 4.11. Experimental versus cross-validated activity for the amino acid residues.

These results can be compared to the previous work of Damborský, who studied the same data set [214]. The model proposed by this author used 402 descriptors from the AAindex database [215-217], which were optimally combined to generate satisfactory data estimations. The best model yielded r^2=0.859 and q^2=0.809 using only 2 descriptors. In these calculations, a mechanistic interpretation of the model was discussed. The parameters employed were the principal component IV connected to property A362 in the database [218] and the low and medium-range interaction energy, coincident with property P214 [219]. Principal component IV, derived from Sneath's PCA analysis, was interpreted by the author as a hydroxythiolation effect, involving amino acid residues with hydroxyl and sulphydryl groups. The second descriptor, namely the low and medium-range interaction energy for the amino acid residues, is the sum of the non-bonded interaction energy between the central chain and the side-chains, corresponding to a low-range effect; and between two residues located within ten residues along the chain, associated to a medium-range effect, calculated using the Lennard-Jones potential. The Lennard-Jones potential is the field used in CoMFA approach to represent the molecular steric effects, and hence the satisfactory results derived with MQSM are justified. Moreover, Wells et al [220] suggested in another paper that the subtilisin mutant activity was mainly related to the size and branching of the substituent residues, features that may be included in the information contained into the molecular density functions. Quantum Similarity methodology does not provide so accurate results, but on the other hand they do not depend on the number of descriptors of the database chosen. Some caution is necessary when comparing the results, since Damborský's data have not been scaled. This fact confers a low confidence to the model, because of the problematic influence of the data distribution already discussed. Regardless of some exceptions, the structure-function relationship for this molecular set is quite clear: amino acid mutations possessing rings {His, Phe, Trp, Tyr} have a low activity, from 4 to 4.9 units; whereas amino acids with sulphur atoms {Cys, Met} have a very high activity: 138 and 100 units, respectively. Almost all the amino acid residues with small side-chains {Ala, Asn, Gly, Leu, Ser, Thr, Val} present intermediate-high activities (10<K<60), and those with long side-chains {Arg, Gln, Glu, Lys} possess an intermediate-low action, lower than 10 units. This association, based on shape considerations, seems that can be elucidated with techniques employing Quantum Similarity concepts.

Finally, the randomization test was applied to confirm that the obtained relationships were not due to an overparametrization of the model. Thus, 100 regressions were built using randomized responses, under the optimal conditions (MPVM, 3 PCs). The results are shown in figure 4.12. The actual activity yields the best results, and none of the altered responses generate a valuable predictive model. All the models with unreal activities yield q^2 values less than 0.5, and many of them are even negative. These results indicate that a real structure-function has been achieved, and that the effect of single-point mutations on the function of the whole enzyme can be quantified using MQSM. These results can be put together with other two previous studies, where QSFR models were built for haloalkane dehalogenase [221] and α-subunits of tryptophan synthase [222] enzymes using quantum similarity matrices and a similar protocol as the one used here.

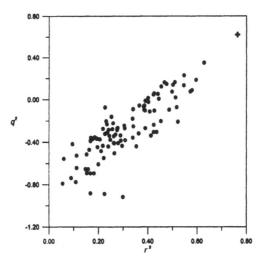

Fig. 4.12. Randomization test for the aminoacids. Permuted activities are marked with circles, and the correct activity with a cross.

5 Quantum self-similarity measures as QSAR descriptors

Sometimes, drug-receptor interactions are simple enough to be accurately characterized by a single parameter linear relationship. This was a general fact in early QSAR models, in which descriptors such as log P or Hammett σ were used as sole parameters. In this kind of systems, therefore, it is not necessary to use such a sophisticated QSAR approach as detailed in chapter 4. A simpler method can be constructed by neglecting the off-diagonal similarity matrix while using only the diagonal elements, constituting the so-called *quantum self-similarity measures* (QS-SM). This simplification avoids the problem of selecting a molecular alignment, because the compared electron distributions belong to the same molecule, so the alignment is irrelevant. This new approach also permits, after a subsequent manipulation, the treatment of molecular fragments within the quantum similarity framework.

The present chapter is organized as follows: first, a connection of the quantitative structure-property relationships (QSPR) with Quantum Similarity is proposed. Then, the correlation of QS-SM and usual classical QSAR descriptors is evidenced, and finally, these measures are applied to find out simple mathematical equations to describe biological activities.

5.1
Simple QSPR based on QS-SM

It was previously discussed in section 3.1.1, how the expectation value computation formalism within the framework of Quantum Mechanics can be expressed in a discrete form, involving MQSM. Equation 3.2 constitutes the theoretical basis of QSPR, connecting observable quantities: physico-chemical properties or biological activities, with theoretical descriptors, namely MQSM:

$$y_I = \langle \Omega \rangle = \int \Omega(\mathbf{r}) \rho_I(\mathbf{r}) d\mathbf{r} \approx \boldsymbol{\beta}^T \mathbf{z}_I \qquad (5.1)$$

Although equation 5.1 represents a general theoretical form of QSPR, in some cases it is also useful to obtain a simpler form. This new expression can be closely related to the well-known linear free-energy relationship (LFER) equation. In order to introduce this simplification, equation 5.1 can be rewritten by dividing the self-similarity part from the off-diagonal terms:

$$y_I \approx \boldsymbol{\beta}^T \mathbf{z}_I = \beta_I Z_{II} + \sum_{J \neq I} \beta_J Z_{IJ} \qquad (5.2)$$

When dealing with some homogeneous series of compounds, there can be assumed that the off-diagonal MQSM contribution terms are roughly constant, yielding:

$$y_I \approx \beta Z_{II} + \alpha \qquad (5.3)$$

Thus, in this approach only the independent term α and the slope β need to be computed. As a result, the simplified equation expresses a straightforward linear relationship between the property of a molecule and its QS-SM. The following sections will provide some application examples of the use of QS-SM as unique molecular descriptors.

5.2
Characterization of classical 2D QSAR descriptors using QS-SM

The considerable success of MQSM in QSAR analysis has suggested the possibility of connecting them with some widely used classical QSAR descriptors. Parameters such as the octanol-water partition coefficient, log P, or Hammett's σ constant are parameters often employed in Hansch-type QSAR studies [223], which have been used to describe the properties of hundreds of molecular families. In this section, QS-SM appears to be related to both parameters, with promising results. This seems to indicate that, for homogeneous molecular series, the information contained in the classical descriptors can be perfectly reproduced by parameters coming from a quantum-mechanical basis, and more precisely by using simply QS-SM.

5.2.1
QS-SM as an alternative to log P values

The high complexity in drug-receptor interactions makes in many cases impossible to determine empirically the precise substitutions necessary to improve the existing compounds. Thus, an important step in the design of new active drugs constitutes the QSAR analyses, which try to correlate experimental biological data with theoretical descriptors related to the structures of the molecules involved. Among the descriptors most frequently used, the octanol-water partition coefficient, log P, has been proved to have a powerful capacity of description. Log P may be interpreted, as a measure of the hydrophobic character of a molecule, and it seems commonly accepted that some of the relevant drug-receptor interactions are highly determined by the hydrophobicity of the ligands. Hence, there is no doubt about its success as a molecular descriptor.

Although such a parameter is generally available from experimental assays for a large series of molecules, some theoretical approaches have been proposed to estimate its value, mainly based on empirical rules. The log P computation techniques rely on additive atomic group contributions. Several studies [223-235] attempted to describe log P using different approaches, for example: Moriguchi et al reported a comparative study between these methodologies [236]. An alternative possibility consists of using QS-SM, where the interplay between the hydrophobic and lipophilic affinities of the molecule is included in the comparison of calculated electron distributions of the same molecular structure computed in two solvent environments: water and 1-octanol [237,238]. In this paper, the octanol-water partition coefficient was mimicked by considering such a QS-SM calculation, involving the product of two densities corresponding to the same molecule, but each one with the molecular structure immersed in a different solvent: namely water and 1-octanol. Later, it was proved that employing only gas-phase QS-SM, satisfactory correlations were also achieved [239]. Thus, this simpler approach will be used here.

An initial idea related to the current use of QS-SM as single descriptors arises from the proposal of Klopman et al. [227,228], who developed a method based on hydrophobic atomic contributions, connecting, in this way, log P to the electrostatic interactions, involving point-like charge densities. However, here in the quantum similarity context, instead of using discrete point charges on individual atoms, a continuous density function is chosen to characterize the charge distribution in the molecule.

As an initial simple illustrative example, the correlation between overlap QS-SM and log P was searched for a set of 7 ethers. The structure of the ethers was constructed simply by adding methylene groups to the initial $C_2H_5OC_2H_5$ molecule. Such a homogeneous set constitutes a good candidate for correlating its properties with QS-SM. Table 5.1 gathers the studied molecular set, the computed QS-SM and the log P values [240]. Ether geometries were optimized with the semiempirical AM1 Hamiltonian [241] using AMPAC program [242]. Molecular density functions were constructed with the promolecular ASA fitting, using valence electrons modified by AMPAC atomic charges.

An excellent correlation between overlap QS-SM and log P exists for this series, as shows the correlation coefficient $r^2=1.000$. This correlation is represented in fig. 5.1. The equation relating both magnitudes is:

$$Z_{II} = 83.8979 + 27.7006 \log P \qquad (5.4)$$

Correlations obtained using Coulomb QS-SM were also very satisfactory, yielding: $r^2=0.996$. These results complement those reported in previous studies [237,239].

Table 5.1. Ethers, QS-SM and log P values.

Ethers	Overlap QS-SM	log P
$C_2H_5OC_2H_5$	109.91	0.89
$C_3H_7OC_3H_7$	140.53	2.03
$C_4H_9OC_4H_9$	171.09	3.21
$C_5H_{11}OC_5H_{11}$	201.70	4.29
$C_6H_{13}OC_6H_{13}$	232.30	5.37
$C_7H_{15}OC_7H_{15}$	262.90	6.45
$C_8H_{17}OC_8H_{17}$	293.50	7.53

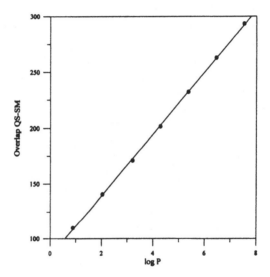

Figure 5.1. Relationship between QS-SM and log P for the 7 ethers. Solid line corresponds to a linear regression.

5.2.2
QS-SM as an alternative to Hammett σ constant

In the late 30's, Hammett verified that the ionization equilibrium constants of the *meta* and *para* substituted benzoic acids were correlated. This led him to define a constant that somehow was able to characterize the electronic effects produced by the substituents. This constant, denoted by the Greek letter σ, became a parameter possessing the power to describe the activity of several molecular families [243]. The intention of the present study is to show that QS-SM of appropriately selected molecular fragments do indeed describe the variation of substituent effect in a given reaction series, so they can be regarded, in this point of view, as equivalent to the Hammett's σ constant. A first study about this matter has been already reported, where an excellent linear correlation between Hammett's σ and QS-SM was found for five series of aromatic compounds [238]. The case example presented here is an extension of the previous reference study, in which another series of acids was studied. The methodology employed follows the theoretical framework explained above. Even if, in principle, QS-SM corresponding to the whole molecules can be used in the correlations with σ, it has been found useful not to characterize the whole molecule with the associated Z_{ll}, but only certain selected intramolecular atomic fragments. These molecular fragments can be, in turn, identified as the molecular parts that actively participate in the reaction process. The philosophy underlying this assumption is based on the fact that the reaction center is usually the region most strongly involved in the reactive process. In this sense, neglecting any contaminating interaction with the remaining part of the molecule may result in a sensitivity increment of the corresponding descriptor to any external effects. Such a specific limitation to a particular molecular fragment is made possible only when the reaction center is unambiguously known. Here, a series of 12 carboxylic acids is analyzed. The general structure for this series is shown in figure 5.2. As in the series previously studied by Ponec et al [238], the active part of the compounds can be clearly assigned to the COOH group. As a result, QS-SM will be calculated only for this fragment, and variations on their values will be associated with differences in the type of the remaining substituent X. The remaining substructure, namely the benzene ring plus the $(CH_2)_2$ chain, are considered contaminating fragments that may produce "noise" in the sensitivity of the reaction center.

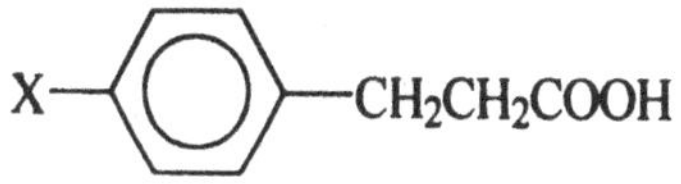

$$X-\langle\bigcirc\rangle-CH_2CH_2COOH$$

Figure 5.2. Structure of the carboxylic acids studied. X indicates the variable substituent.

If either Hammett's σ or QS-SM are considered to linearly describe the dissociation equilibrium, this means that the following equation will hold:

$$pK_X = a\,\sigma_X + b = a'\,Z_{XX}^{COOH} + b'$$
$$\Rightarrow Z_{XX}^{COOH} = \alpha\,\sigma_X + \beta \tag{5.5}$$

being X the variable substituent. So, in this particular study, Hammett's σ will be linearly related to fragment QS-SM of COOH group.

The calculations for all the participating molecules have been carried out using the semiempirical AM1 Hamiltonian [241] using the AMPAC package [242]. As in the previous case, molecular density functions were described using promolecular ASA, whose coefficients were adapted by valence electrons modified by AMPAC atomic charges. In all cases, the structures of the substituted acids have been completely optimized and the resulting geometries and charge distributions have been used as the input data for the subsequent QS-SM calculations. Overlap QS-SM was used as molecular descriptor. Table 5.2 shows the 12 different substituents considered and the values of fragmental QS-SM and Hammett's σ for this particular substituent.

The adjustment between the Hammett σ and QS-SM is carried out by means of a least squares procedure. The computed equation, relating both descriptors is:

$$Z_{XX}^{COOH} = -0.10338\sigma_X + 113.311 \qquad (5.6)$$

Table 5.2. Carboxylic acids. Variable substituents, COOH fragment QS-SM and Hammett's σ.

X	Z_{XX}^{COOH}	σ
Br	113.285	0.22
H	113.327	0
CH$_3$	113.333	-0.14
CH$_3$O	113.329	-0.28
Cl	113.293	0.22
CN	113.260	0.71
F	113.291	0.06
NO$_2$	113.207	0.81
CH$_3$CH$_2$	113.336	-0.13
N(CH$_3$)$_2$	113.371	-0.63
CCl$_3$	113.268	0.46
CF$_3$	113.248	0.53

A satisfactory linear relationship exists, as assesses the determination coefficient: $r^2 = 0.921$. These results, together with those reported previously [238] justify the assumption that QS-SM can represent satisfactorily the role of Hammett's σ in QSAR studies dealing with homogenous molecular series. Figure 5.3 depicts the observed correlation between Coulomb QS-SM and Hammett's σ constant.

As a conclusion, the above results prove that the slight changes on the electron distribution, produced by the modification of a given substituent for a homogeneous molecular set, can be detected and quantified with QS-SM. Due to the fact, assessed by a great amount of LFER studies, that substituent effect descriptors are able to describe satisfactorily physico-chemical properties and even biological activities, QS-SM could be used in the same way. This point is discussed in the next section.

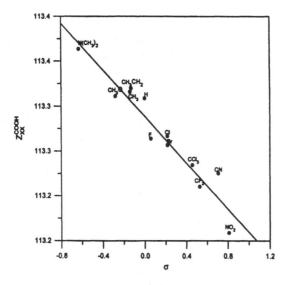

Figure 5.3. Dependence of calculated QS-SM for the series of substituted carboxylic acids on the Hammett's σ constant. The solid line is the optimal fitting of the data.

5.3
Description of biological activities using fragment QS-SM

Going further in the practical use of QS-SM, it must be stressed that they can be straightforwardly applied in QSAR analyses. The methodology presented here is mainly based on the utilization of fragment QS-SM as molecular descriptors in the generation of QSAR models. This approach is founded in the aforementioned previous studies [237-239,244], where it was demonstrated that QS-SM of the whole molecule or of appropriate selected fragments can be employed as an alternative to some classical physico-chemical descriptors, such as log P and the Hammett's σ constant. This fact was evidenced in the precedent example, in which QS-SM were computed only for the COOH fragment, using the fact that it was the obvious reaction center for carboxylic acids. Nevertheless, in the major part of QSAR problems related to biological activity or toxicity, the sensitive molecular sites responsible for activity are not so evident. The objective of the present study is to propose a general methodology capable to identify those molecular fragments that best describe a molecular property, without any imposed restriction or *a priori* specification. Basically, the procedure allows for the detection of those molecular regions, common to all the molecular series, which are responsible for a high biological response. This leads to obtain an active region pattern, of evident interest for drug design purposes. A necessary condition for the application of the present methodology is that the studied compounds have to exhibit some common structural features. The main steps of the proposed procedure are: 1) Generate all the possible molecular fragments. The present algorithm permits to compute any molecular fragment. However, in this case only 1, 2 and 3 bonded atoms groups have been considered. 2) Compute all QSAR models using previous fragment QS-SM as molecular descriptors. Select all those QSAR models accounting for $q^2 > 0.6$, the threshold for the prediction coefficient to be considered as satisfactory. 3) Represent in a histogram the frequency of the backbone atoms appearing in the fragment QS-SM of the selected QSAR models. The atoms with a higher appearance will point out their greater relevance for the property description.

As an application example of this procedure, the antibacterial activity for a series of quinolones has been studied and the molecular regions relevant for activity were elucidated. Nalidixic acid was one of the first quinolones discovered to possess antibacterial activity [245]. These older drugs were of relatively minor significance, because of their limited therapeutic utility and the rapid development of bacterial resistance. Later, a new generation of quinolones was developed, including norfloxacin [246], oxolinic acid, pipemidic acid and flumequine as potent, orally active antibacterials [247]. In addition, few side effects appeared to accompany the use of these quinolones, and microbial resistance to their action did not develop rapidly [248,249]. The success of this new class of chemotherapeutic agents led to a great effort of synthesis of new quinolone derivatives [250-255]. Among these novel compounds, derivatives of 7-(3'-amino-3'-methyl-1'-azetidinyl) quinolones were designed, synthesized and tested, assessing a broad-spectrum activity against Gram-positive organisms [256]. This set, made up of 29 quinolones, was studied using Quantum Similarity techniques in order to correlate the molecular structure with the antibacterial activity. Table 5.3 shows the general structure for the quinolone set, in which all atoms are numbered. These labels will serve to identify the molecular groups employed to construct the regression models. In addition, a table with the substituents defining each compound is given, as well as the antibacterial activity for three types of Gram-positive bacteria [256]. Activity data were expressed in terms of the minus logarithm, in order to reduce scale differences. However, activity data present a particular distribution that makes difficult the quantitative analysis, due to the fact that the property exhibits a low variation and possesses almost discrete values.

Let m be the number of existing molecular fragments and k the number of effective descriptors involved in the QSAR model. In order to select the optimal fragment QS-SM, all possible combinations of m fragments with k descriptors were generated using a *nested summation symbol* algorithm [257,258]. Although the optimal QSAR model corresponds to the one with the highest q^2, all models leading to $q^2 > 0.6$ were taken into consideration. In this way, one can get a precise idea of which are the regions relevant for activity, and then construct the active region pattern. Thus, the results presented here may be considered more qualitative than quantitative.

Geometries for the 29 quinolones were optimized using semiempirical AM1 Hamiltonian [241] with AMPAC 6.0 program [242]. As in the previous cases, ASA density functions were constructed using valence electrons modified by AMPAC atomic charges. The common structural features for antibacterial quinolones are composed by 28 atoms, which are the labeled atoms in Table 5.3. The first step of the procedure consists of defining all possible groups of 1, 2 and 3 atoms. For the groups of one atom, large errors would be present if hydrogen atoms were considered as a single fragment. QS-SM for hydrogen atom is negligible compared to QS-SM of C, N, O and F atoms [259]. In addition, H atoms are not very well described in the current used ASA, because only one spherical function was used for them. Thus, hydrogen atoms considered individually as a single fragment could produce considerable errors, and therefore only C, N, O and F atoms as single-atom groups were included. Consequently, only the remaining 19 non-hydrogen single-fragment QS-SM were computed.

Alternatively, H atoms were taken into account for the generation of molecular groups of 2 and 3 atoms. 29 different pairs of bonded atoms were defined, as well as a number of 48 triads of correlative atoms. As a result, a total number of 96 molecular groups were generated. Due to the fact that both overlap and Coulomb QS-SM were computed, a gross total number of $m=192$ molecular descriptors were considered in the calculation. Overlap and Coulomb QS-SM values will be denoted with the subindex OVE and COU, respectively. The predictive models were chosen to have 4 parameters. In order to avoid scaling differences between the values of the fragment QS-SM, every descriptor was self-scaled by standardizing its values in the usual way: by subtracting the arithmetic mean and dividing by the standard deviation. This yields standardized descriptor columns having mean zero and standard deviation one. As a consequence, the estimators of the multilinear regression have comparable values.

For each one of the different bacteriological systems studied, two histograms are displayed, corresponding to the overlap and Coulomb fragment QS-SM. These histograms represent the frequency of appearance of the different atoms of the common skeleton in the QSAR models with $q^2>0.6$. This separate representation may evidence differences between both types of measures. It is accepted that overlap QSM have a strong shape character, whereas Coulomb-like provide an electrostatic picture on the structure-activity relationships. In addition, the smoother shape of the density functions weighted by the Coulomb operator may be interpreted as also providing a volume, steric effect.

Table 5.3. General structure, substituents and *in vitro* antibacterial activity (MIC, μg/ml) of 7-(3'-amino-3'-methyl-1'-azetidinyl) quinolones.

compd	A	R_1	R_2	R_3	R_5	Bc	Sf	Sa
31i	CH	c-C$_3$H$_5$	H	OH	H	0.921	0.301	0.921
32i	CF	c-C$_3$H$_5$	H	OH	H	0.921	0.301	0.921
33i	CF	c-C$_3$H$_5$	H	OH	NH$_2$	1.222	0.602	1.222
34i	CCl	c-C$_3$H$_5$	H	OH	H	1.222	0.602	1.222
35i	N	c-C$_3$H$_5$	H	OH	H	0.921	0.602	0.602
36i	CH	c-C$_3$H$_5$	–SNH–		H	1.523	0.921	1.222
37i	CF	c-C$_3$H$_5$	–SNH–		H	0.921	0.301	0.921
38i	CH	C$_2$H$_5$	H	OH	H	0.000	0.602	0.301
39i	CF	C$_2$H$_5$	H	OH	H	0.301	-0.301	0.301
40i	CF	C$_2$H$_5$	H	OH	NH$_2$	0.000	0.000	0.301
41i	CCl	C$_2$H$_5$	H	OH	H	0.602	0.000	0.602
42i	N	C$_2$H$_5$	H	OH	H	0.000	-0.301	0.000
43i	CH	2,4-F$_2$Ph	H	OH	H	0.602	0.921	0.602
44i	CF	2,4-F$_2$Ph	H	OH	H	0.301	0.602	0.602
45i	CF	2,4-F$_2$Ph	H	OH	NH$_2$	1.222	1.523	1.222
46i	CCl	2,4-F$_2$Ph	H	OH	H	1.222	0.602	1.222
47i	N	2,4-F$_2$Ph	H	OH	H	0.921	0.301	0.921
48i	CH	CH$_2$CH$_2$F	H	OH	H	0.000	-0.301	0.000
49i	CF	CH$_2$CH$_2$F	H	OH	H	0.000	-0.602	0.000
50i	CCl	CH$_2$CH$_2$F	H	OH	H	0.301	0.000	0.602
51i	N	CH$_2$CH$_2$F	H	OH	H	0.301	-0.301	0.301
52i	CH	t-C$_4$H$_9$	H	OH	H	0.602	0.301	0.602
53i	CF	t-C$_4$H$_9$	H	OH	H	0.301	-0.301	0.301
54i	N	t-C$_4$H$_9$	H	OH	H	0.921	0.000	0.921
55i	CH	4-FPh	H	OH	H	0.921	-0.301	0.921
56i	CF	4-FPh	H	OH	H	0.921	-0.301	0.921
57i	CCl	4-FPh	H	OH	H	1.222	0.602	1.222
58i	N	4-FPh	H	OH	H	0.602	-0.602	0.602
59i	COCH$_2$CH(CH$_3$)	H	OH	H	0.301	0.301	0.301	

Bc: *Bacillus cereus* ATCC 11778; Sf: *Streptococcus faecalis* ATCC 10541; Sa: *Staphylococcus aureus* ATCC 25178.

5.3.1
Activity against *Bacillus cereus* ATCC 11778 (Bc)

The first property analyzed is denoted by Bc, and corresponds to the activity against bacterium *Bacillus cereus* ATCC 11778. For this bacteria, were obtained a total of 4062 QSAR models having $q^2 > 0.6$. From all these models, a few ones yielding the most significant results have been selected, and shown as follows:

$$\log(\mathcal{V}_{MIC}) = 0.536\, Z_{OVE}^{C(7)N(1')C(4')} - 0.512\, Z_{COU}^{N(1)C(9)} + 0.364\, Z_{COU}^{C(3')N(19)\,H(28)}$$
$$- 0.165\, Z_{COU}^{C(18)H(24)H(25)} + 0.663 \tag{5.7}$$

$$r^2 = 0.771; \quad q^2 = 0.682$$

$$\log(\mathcal{V}_{MIC}) = 0.506\, Z_{OVE}^{C(7)N(1')} - 0.499\, Z_{COU}^{N(1)C(9)} + 0.362\, Z_{COU}^{C(3')N(19)\,H(28)}$$
$$- 0.150\, Z_{COU}^{C(18)H(24)\,H(25)} + 0.663 \tag{5.8}$$

$$r^2 = 0.770; \quad q^2 = 0.681$$

$$\log(\mathcal{V}_{MIC}) = -0.468\, Z_{OVE}^{N(1)C(9)} + 0.525\, Z_{OVE}^{C(7)N(1')C(2')} + 0.561\, Z_{OVE}^{C(3')N(19)\,H(28)}$$
$$- 0.378\, Z_{COU}^{N(19)H(27)} + 0.663 \tag{5.9}$$

$$r^2 = 0.776; \quad q^2 = 0.678$$

$$\log(\mathcal{V}_{MIC}) = -0.468\, Z_{OVE}^{N(1)C(9)} + 0.563\, Z_{OVE}^{C(3')N(19)} + 0.525\, Z_{OVE}^{C(7)N(1')C(2')}$$
$$- 0.381\, Z_{COU}^{N(19)H(27)} + 0.663 \tag{5.10}$$

$$r^2 = 0.776; \quad q^2 = 0.678$$

The main characteristic of the previous equations consists of the molecular fragments involved in the models are basically the same. Variations appear only for one or two atoms. This fact enhances the robustness of the model, since the fragments that best describe the biological activity are well localized. Another remarkable fact observed in the QSAR equations for Bc bacteria is that, for some particular fragments, both overlap and Coulomb measures can be interchanged without relevant variations in the statistical results. A clear example of this behavior is given by the fragment made up of atoms N(1)C(9). This permits to conclude that both types of QS-SM, for certain fragments, provide almost the same information. Figure 5.4 shows the histograms for each QS-SM. The atoms have been labeled according to the figure shown in Table 5.3.

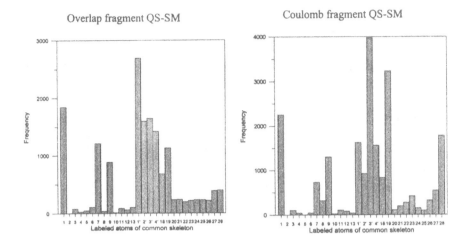

Figure 5.4. Histograms for overlap and Coulomb fragment QS-SM. Atoms are labeled according to table 5.3.

From here it can be seen that both overlap and Coulomb QS-SM provide the same kind of information. The atoms that participate in a greater extent within the models are: N(1), C(9) and C(7), which are the atoms closer to substituent A. In particular, N(1) and C(9) constitute part of the quinolone ring. N(1'), C(2'), C(3'), C(4'), atoms that form the azetidinydil ring; and N(19), H(27) and (H28), which form the amino group.

Depicted histograms agree with the QSAR equations. Atoms marked in the plot coincide with the QS-SM fragments that appear in the most predictive models. This fact enhances the use of histograms as a useful tool to extract information from the generated QSAR model. Moreover, this allows plotting a kind of pharmacophore, where the selected fragments are marked. This is shown in Figure 5.5.

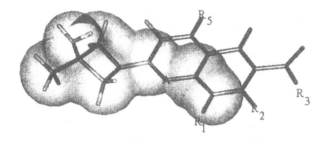

Figure 5.5. Proposed active regions for the action of quinolones on Bc bacteria.

5.3.2
Activity against *Streptococcus faecalis* ATCC 10541 (Sf)

The second property analyzed is denoted by Sf, and corresponds to the activity against bacterium *Streptococcus faecalis* ATCC 10541. In this case, only 38 models led to $q^2 > 0.6$ values, and therefore the equations are not so much reliable. QSAR models with the highest q^2 values for this property are the following:

$$\log(V_{MIC}) = 0.070 \, Z_{OVE}^{O(12)} - 0.376 \, Z_{COU}^{N(1)C(9)} - 1.042 \, Z_{COU}^{C(18)H(24)}$$
$$+ 0.803 \, Z_{COU}^{C(18)H(26)} + 0.209 \tag{5.11}$$
$$r^2 = 0.747; \quad q^2 = 0.625$$

$$\log(V_{MIC}) = 0.070 \, Z_{COU}^{O(12)} - 0.376 \, Z_{COU}^{N(1)C(9)} - 1.042 \, Z_{COU}^{C(18)H(24)}$$
$$+ 0.803 \, Z_{COU}^{C(18)H(26)} + 0.209 \tag{5.12}$$
$$r^2 = 0.747; \quad q^2 = 0.625$$

$$\log(V_{MIC}) = 0.068 \, Z_{OVE}^{C(11)O(12)} - 0.375 \, Z_{COU}^{N(1)C(9)} - 1.043 \, Z_{COU}^{C(18)H(24)}$$
$$+ 0.803 \, Z_{COU}^{C(18)H(26)} + 0.209 \tag{5.13}$$
$$r^2 = 0.746; \quad q^2 = 0.622$$

$$\log(V_{MIC}) = 0.064 \, Z_{COU}^{C(11)O(12)} - 0.372 \, Z_{COU}^{N(1)C(9)} - 1.043 \, Z_{COU}^{C(18)H(24)}$$
$$+ 0.803 \, Z_{COU}^{C(18)H(26)} + 0.209 \tag{5.14}$$
$$r^2 = 0.745; \quad q^2 = 0.619$$

$$\log(V_{MIC}) = 0.346 \, Z_{COU}^{C(2)C(3)} - 1.010 \, Z_{COU}^{C(18)H(24)} + 0.783 \, Z_{COU}^{C(18)H(26)}$$
$$- 0.503 \, Z_{COU}^{N(1)C(2)C(9)} + 0.209 \tag{5.15}$$

$$r^2 = 0.748; \quad q^2 = 0.617$$

$$\log\left(\chi_{MIC}\right) = -0.352\, Z_{COU}^{N(1)C(9)} - 1.175\, Z_{COU}^{C(18)H(24)} + 0.735\, Z_{COU}^{C(18)H(26)}$$
$$+ 0.219\, Z_{COU}^{C(18)H(24)H(25)} + 0.209 \tag{5.16}$$

$$r^2 = 0.751; \quad q^2 = 0.610$$

The first difference which can be appreciated, when comparing this case with the previous one, consists of q^2 values are slightly lower. Another aspect, already discussed for the equations for *Bacillus cereus* bacteria, is that for some fragment QS-SM, overlap and Coulomb QS-SM can be used without distinction yielding to the same statistical results. For instance, the only difference between QSAR models 5.11 and 5.12 is the use of $Z_{OVE}^{O(12)}$ instead of $Z_{COU}^{O(12)}$. The same occurs in equations 5.13 and 5.14 with respect of the C(11)O(12) group. On the other hand, due to the fact that the remaining fragments correspond to Coulomb-like measures and that they remain constant in the models, this indicates that overlap and Coulomb explain, in general, different properties of the same molecular fragments. This can be assessed exchanging such fragment QS-SM by their corresponding overlap homologue. For instance, only by changing $Z_{COU}^{C(18)H(24)H(25)}$ by $Z_{OVE}^{C(18)H(24)H(25)}$ in equation 5.16, the q^2 value decreases to 0.584. A similar behavior is observed in equation 5.11, since if $Z_{COU}^{N(1)C(6)}$ is substituted by $Z_{OVE}^{N(1)C(6)}$, the predictive capacity is reduced to $q^2 = 0.597$. Furthermore, in equation 5.16 appear three descriptors related to the group C(18)H$_3$, all of them associated to Coulomb measures. If these descriptors are exchanged by their corresponding overlap homologues, q^2 decreases to 0.531. This could indicate that the influence of the methyl group in the activity might be attached to a steric effect, which is well characterized by the Coulomb QS-SM.

Figure 5.6 shows the histograms for overlap and Coulomb fragment QS-SM. They represent the frequency of appearance of the common skeleton atoms in models with $q^2 > 0.6$.

Overlap fragment QS-SM   Coulomb fragment QS-SM

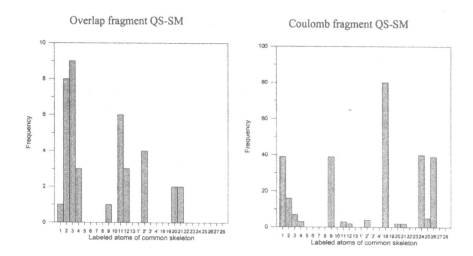

Figure 5.6. Histograms for overlap and Coulomb fragment QS-SM. Atoms are labeled according to table 5.3.

Some differences can be appreciated in the overlap and Coulomb histograms here. Overlap QS-SM selected C(2)C(3)C(4), C(11)O(12) fragments and the group C(2')H$_2$. On the other hand, the most frequent regions according to Coulomb description were the N(1)C(9) fragment and the C(18)H$_3$ group. The main difference with the previous example was the inclusion of the carboxylic acid moiety at position 3, while atoms of azetidinydil ring atoms were rejected. Starting from these histograms, the most relevant fragments for Sf activity can be marked in a plot, as shows in figure 5.7.

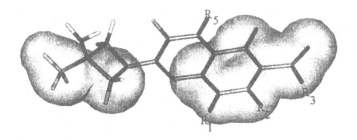

Figure 5.7. Proposed active regions for the action of quinolones on Sf bacteria.

5.3.3
Activity against *Staphylococcus aureus* ATCC 25178 (Sa)

Finally, Sa, the activity against bacterium *Staphylococcus aureus* ATCC 25178 was studied. *Staphylococcus aureus* is of a great medical interest because it intervenes in diseases such as abscesses, bacteremia, endocarditis, pneumonia, meningitis and osteomyelitis, among others [260]. QSAR models with highest predictivity for these bacteria are:

$$\log(\textstyle\frac{1}{MIC}) = -0.359\,Z_{OVE}^{N(1)} + 0.166\,Z_{COU}^{C(4)O(10)} + 0.396\,Z_{COU}^{N(1')C(4')H(22)}$$
$$+\, 0.298\,Z_{COU}^{C(3')C(4')N(19)} + 0.683 \tag{5.17}$$
$$r^2{=}0.844;\ q^2{=}0.754$$

$$\log(\textstyle\frac{1}{MIC}) = -0.359\,Z_{COU}^{N(1)} + 0.166\,Z_{COU}^{C(4)O(10)} + 0.396\,Z_{COU}^{N(1')C(4')H(22)}$$
$$+\, 0.298\,Z_{COU}^{C(3')C(4')N(19)} + 0.683 \tag{5.18}$$
$$r^2{=}0.844;\ q^2{=}0.754$$

$$\log(\textstyle\frac{1}{MIC}) = -0.361\,Z_{OVE}^{N(1)} + 0.373\,Z_{OVE}^{N(1')C(4')H(22)} + 0.168\,Z_{COU}^{C(4)O(10)}$$
$$+\, 0.268\,Z_{COU}^{C(3')C(4')N(19)} + 0.683 \tag{5.19}$$
$$r^2{=}0.846;\ q^2{=}0.753$$

$$\log(\textstyle\frac{1}{MIC}) = 0.373\,Z_{OVE}^{N(1')C(4')H(22)} - 0.361\,Z_{COU}^{N(1)} + 0.168\,Z_{COU}^{C(4)O(10)}$$
$$+\, 0.268\,Z_{COU}^{C(3')C(4')N(19)} + 0.683 \tag{5.20}$$
$$r^2{=}0.846;\ q^2{=}0.753$$

These correlations are better than the previous models designed for the two already studied bacteria. A huge amount of satisfactory models, 95893, were generated, all of them accounting for $q^2{>}0.6$. Results for *Staphylococcus aureus* and *Bacillus cereus* are quite similar. Azetidinydil ring appears to be one of the most important fragments to describe the quinolone activity. In addition, as it has been stated in the previous cases, N(1) from the quinolone ring plays an essential role in the model. The most significant difference with the first case example is the appearance of the C(4)O(10) fragment at position 4 of the quinolone ring.

From the histogram plots (Figure 5.8) can be observed that the most frequent atoms coincide with those constituting the most predictive equations. The only divergence appears for the C(4)O(10) group, which is not specially highlighted in the figure.

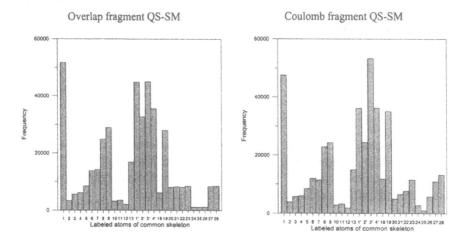

Figure 5.8. Histograms for overlap and Coulomb fragment QS-SM. Atoms are labeled according to table 5.3.

As in the previous cases, the active regions proposed can be drawn, according to the information contained in the histograms and in the optimal equation. The molecular regions relevant for Sa activity are given in Figure 5.9.

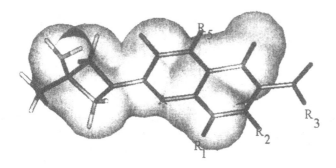

Figure 5.9. Proposed active regions for the action of quinolones on Sa bacteria.

6 Electron-electron repulsion energy as a QSAR descriptor

In this chapter, the expectation value of the interelectronic repulsion energy operator, $<V_{ee}>$, is presented as a kind of QS-SM, which consequently can be used as a molecular descriptor in QSAR applications. The efficiency of this parameter in QSAR for different molecular sets will be here examined.

6.1 Connection between the electron-electron repulsion energy and QS-SM

In section 2.6, the general formulation of MQSM was exposed. Assuming that both compared molecules are the same and that they are located in this way at the same position, then a quantum self-similarity measure (QS-SM) can be defined as the integral:

$$Z_{AA}(\Omega) = \int \Omega(\mathbf{r}) \rho_A^2(\mathbf{r}) \, d\mathbf{r} \, . \tag{6.1}$$

This expression is valid for any non-differential operator. Equation (6.1) can be generally extended to multiple QS-SM as:

$$Z_{AA}^{(n)}(\omega) = \int \omega(\mathbf{R}) \rho_A^n(\mathbf{R}) \, d\mathbf{R} \, , \tag{6.2}$$

where $\mathbf{R} = (\mathbf{r}_1, \mathbf{r}_2, ..., \mathbf{r}_n)$ represents the set of particles coordinates. A final step can be performed taking into account the most simple case, $n=1$, yielding the expression:

$$Z_{AA}^{(1)}(\omega) = \int \omega(\mathbf{R}) \rho_A(\mathbf{R}) \, d\mathbf{R} \, . \tag{6.3}$$

This equation represents an alternative formulation of quantum mechanical expectation values for non-differential operators, which can be considered as particular cases of QS-SM, that is:

$$\langle \omega(\mathbf{R}) \rangle = Z_{AA}^{(1)}(\omega) \, . \tag{6.4}$$

Coulomb and exchange operator expectation values, <V_{ee}>, for a p electrons system can be written, at the monoconfigurational ground discrete Hartree-Fock level [261], as:

$$\left\langle V_{ee}\right\rangle = \sum_{i=1}^{p-1}\sum_{j=i+1}^{p}\left[2\left(ii|jj\right)-\left(ij|ij\right)\right],\qquad(6.5)$$

where $(ii|jj)$ and $(ij|ij)$ are the respective Coulomb and Exchange integrals over the molecular orbital (MO) basis set. In this way, the electron-electron repulsion energy can be considered as a sum of self-similarity measures over the occupied MO set [262], which involve the Coulomb operator, defined as:

$$C(\mathbf{R}) = \sum_{i>j}\left|\mathbf{r}_i - \mathbf{r}_j\right|^{-1}\wedge\mathbf{R} = \left(\mathbf{r}_1,\mathbf{r}_2,...,\mathbf{r}_n\right)\cdot\qquad(6.6)$$

The density function of the molecular system can be constructed in the usual way:

$$\rho_A(\mathbf{R}) = \left|\Psi_A(\mathbf{R})\right|^2,\qquad(6.7)$$

where $\Psi_A(\mathbf{R})$ is the approximate wavefunction. So, an equivalent expression to equation (6.5) appears to be:

$$\left\langle V_{ee}\right\rangle = \int C(\mathbf{R})\rho_A(\mathbf{R})\,d\mathbf{R}\,.\qquad(6.8)$$

Equation (6.8) has the same form as equation (6.3), which was in turn deduced from a QS-SM framework. In this manner, it can be affirmed that the evaluation of Coulomb and Exchange operators constitutes a class of QS-SM. Computation of <V_{ee}> is commonly performed by available quantum chemistry software packages, and therefore this new type of MQSM is available to all scientific community.

6.2
<V_{ee}> as a descriptor for simple linear QSAR models

Given a set of molecules, any molecular property can be in principle associated to the expectation value of the interelectronic repulsion energy. When highly homogeneous series are considered, a vector built up from this type of QS-SM can be used as a single molecular descriptor instead of the ($n \times n$) MQSM matrix. Thus, a linear relationship between <V_{ee}> and molecular properties can be imposed:

$$y_A = a\,\left\langle V_{ee}\right\rangle_A + b\,.\qquad(6.9)$$

This equation is a Hansch-type QSAR model, in which the molecular properties are considered to be produced by different effects, which can be represented by different descriptors. In this case, only one descriptor is used. This approach has already been successfully applied in different chemical environments, and thus satisfactory correlations between $<V_{ee}>$ and molecular toxicity have been reported [263], as well as valuable descriptions of physico-chemical properties or biological activities [264].

6.3
Evaluation of molecular properties using $<V_{ee}>$ as a descriptor

Along this section, several molecular examples where $<V_{ee}>$ is used as a molecular descriptor are presented. The information contained in this descriptor is strongly connected to molecular size, but it is also able to reflect electronic effects as well, due to its quantum chemical origin. Thus, $<V_{ee}>$ values discriminate between compounds with the same empirical formula but different connectivity, such as, for instance, heptane and 2,5-dimethylpentane. In this way, it can be stated that $<V_{ee}>$ contains information on the orbital energies differentiating structural isomers and derivatives.

The quality of the models was assessed with the correlation coefficient r^2, the standard deviation of errors of calculation σ_N and the prediction coefficient q^2. In addition, the values of $<V_{ee}>$ have been self-scaled in the statistical sense, that is, by subtracting to all values the arithmetical mean and dividing by the standard deviation. Even though the overall $<V_{ee}>$ values presented in the tables are those obtained directly from calculation program outputs, all correlations have been carried out with the standardized values.

The geometries of all the molecular sets studied were optimized using the AM1 Hamiltonian [265] using AMPAC 6.01 program [266]. The values of $<V_{ee}>$ were computed with a single point calculation over the optimized geometry using Gaussian 94 [267] at a HF/3-21g* computational level.

6.3.1
Inhibition of spore germination by aliphatic alcohols

Cell division and differentiation is a subject of a great interest in Biology. A complete understanding of the whole process is rather difficult, and some certain biological reactions are studied in order to shed light on this phenomenon. For instance, germination of spores may be a useful test process, since it represents a sequential process of intracellular differentiation. In order to obtain information on the mechanism of the initial trigger reaction, structural studies of specific inhibitors are carried out. The effects of several inhibitors on germination and outgrowth of spores of *Bacillus* and *Clostridium* species have been reviewed [268].

In this section, the inhibition of the spore germination of *Bacillus subtilis* by 19 aliphatic alcohols, with both straight and branched chains, are examined. Yasuda-Yasaki et al. [269] demonstrated that low concentrations of alcohols inhibit the germination in L-alanine of *Bacillus subtilis*, and proposed that the inhibition was due to the interaction between a hydrophobic region on the spore and the hydrophobic chain of alcohols. Later they reported a QSAR study relating the hydrophobicity of some alcohols to their inhibitory effect [270]. The biological activity is evaluated here by $-\log I_{50}$, being I_{50} the observed molar concentration of an inhibitor necessary to cause 50% inhibition of germination rate. All data concerning values of $<V_{ee}>$ and biological activity are summarized in Table 6.1.

Table 6.1. Values of $<V_{ee}>$ and biological activity for a set of 19 aliphatic alcohols.

Alcohol	$<V_{ee}>$	$-\log I_{50}$	Alcohol	$<V_{ee}>$	$-\log I_{50}$
Methyl	81.07	0.10	2-Butyl	272.50	1.66
Ethyl	134.84	0.59	Isobutyl	271.06	1.80
n-Propyl	196.33	1.03	*t*-Butyl	277.88	1.19
n-Butyl	263.38	1.77	Isopentyl	348.98	2.31
n-Pentyl	335.28	2.19	Cyclohexyl	433.79	2.31
n-Hexyl	411.32	2.47	2-Octyl	589.36	2.89
n-Heptyl	490.92	3.10	3-Octyl	595.93	2.68
n-Octyl	573.70	3.40	2-Ethyl-1-Hexyl	611.28	3.05
n-Nonyl	659.30	3.46	Phenethyl	550.50	2.11
n-Decyl	747.47	3.52			

The equation relating $<V_{ee}>$ to the inhibition activity of the alcohols, as well as all relevant statistical parameters, are the following:

$$-\log I_{50} = 4.761{\cdot}10^{-3}\langle V_{ee}\rangle + 0.225$$

$$n = 19 \qquad r^2 = 0.863 \qquad \sigma_N = 0.326 \qquad q^2 = 0.828$$

The plot of cross-validated values versus experimental ones plot is presented in Figure 6.1. As it can be seen from the results, a fairly good correlation was obtained using $<V_{ee}>$ as a single molecular descriptor. The correlation coefficient, r^2=0.863, and the value of q^2, 0.828, evidence that a real structure-activity exists.

As it has been commented, Yasuda-Yasaki *et al.* [270] reported a previous study using log P, the hydrophobicity parameter, as a single descriptor for the QSAR model. The model obtained achieved slightly better results than the present ones, yielding r^2=0.920. The aim of the experiments of these authors was to demonstrate that the receptor site possessed a hydrophobic character. In this way, the use of a hydrophobicity descriptor, namely log P, was clearly justified.

However, as indicates the high predictivity of the $<V_{ee}>$ model, molecular descriptors related to electronic effects can also provide a good description of the system, and therefore it can be concluded that the biological activity of aliphatic alcohols can be explained under different points of view.

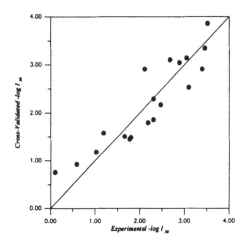

Figure 6.1. Cross-validated versus experimental –log I_{50} for a set of aliphatic alcohols.

6.3.2
Inhibition of microbial growth by aliphatic alcohols and amines

An important topic in environmental chemistry and ecotoxicology consists of the effect of toxic agents on the growth of microbial species. The population growth of protozoa, in particular of ciliates, in varied concentrations of toxic substances has been assessed by comparing a number of specific experimental values, including population density [271], growth rates [272], growth curves [273] and number of generations [274]. In all cases, the most tested species has been *Tetrahymena pyriformis*, a common freshwater hymenostome ciliate, which approximately measures 50 μm in length and 30 μm in width [275]. Modern electronic equipment allows the easy determination of the population growth inhibition, providing a large collection of data for toxicity research. Two different toxicants were studied: a set of 21 linear chain alcohols and a set of 9 aliphatic amines. Testing the effect of different toxic agents on the same organism should allow the elucidation of the reaction mechanism.

The whole molecular set analyzed consists of a mixed series of 30 linear chain alcohols and amines, which inhibit the growth concentration of *Tetrahymena Pyriformis* [276]. The activity was measured by –log GC_{50}, being GC_{50} the 50% inhibitory growth concentration. The values of $<V_{ee}>$ and the biological activity are listed in Table 6.2.

Table 6.2. Values of $<V_{ee}>$ and biological activity for a set of 30 aliphatic alcohols and amines.

Alcohol	$<V_{ee}>$	$-\log GC_{50}$	Alcohol	$<V_{ee}>$	$-\log GC_{50}$
methanol	81.07	-2.77	3-pentanol	350.49	-1.33
ethanol	135.21	-2.41	2-methyl-1-butanol	350.05	-1.13
1-propanol	196.33	-1.84	3-methyl-1-butanol	346.51	-1.13
1-butanol	263.38	-1.52	3-methyl-2-butanol	356.16	-1.08
1-pentanol	335.28	-1.12	(*tert*)pentanol	359.82	-1.27
1-hexanol	411.32	-0.47	(*neo*)pentanol	359.80	-0.96
1-heptanol	490.92	0.02	1-propylamine	187.94	-0.85
1-octanol	573.71	0.50	1-butylamine	254.77	-0.70
1-nonanol	659.30	0.77	1-pentylamine	328.44	-0.61
1-decanol	747.48	1.10	1-hexylamine	405.22	-0.34
1-unidecanol	837.97	1.87	1-heptylamine	485.70	0.10
1-dodecanol	930.62	2.07	1-octylamine	569.45	0.51
1-tridecanol	1121.73	2.28	1-nonylamine	656.12	1.59
2-propanol	200.66	-1.99	1-decylamine	745.42	1.95
2-pentanol	347.32	-1.25	1-unidecylamine	837.10	2.26

The results of the linear relationship between both magnitudes are:

$$-\log GC_{50} = 5.60 \times 10^{-3} \langle V_{ee} \rangle - 2.85$$

$$n = 20 \qquad r^2 = 0.924 \qquad \sigma_N = 0.415 \qquad q^2 = 0.909$$

Figure 6.2 shows the LOO cross-validated $-\log GC_{50}$ versus the experimental values plot.

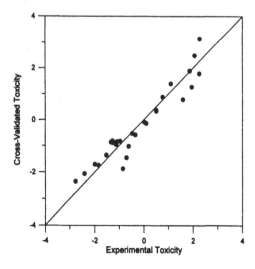

Figure 6.2. Cross-validated versus experimental $-\log GC_{50}$ for the aliphatic alcohols and amines.

The results achieved clearly prove that a fair relationship between $<V_{ee}>$ and -log GC_{50} exists. A high correlation coefficient is obtained, $r^2=0.924$, as well as a notorious predictive coefficient, $q^2=0.909$.

Schultz et al. [276] reported a QSAR study where the same data set was examined. It is commonly accepted that aliphatic compounds act by disrupting physiological processes associated with cellular membranes. The mode of toxic action of these organic compounds, which are both non-ionic and non-reactive, is nonpolar narcosis. In this sense, the response to exposure of animal species to nonpolar narcotics was proposed to be satisfactorily characterized by the octanol-water partition coefficient, log P. Thus, the mentioned authors constructed predictive models using log P as a single descriptor. The correlation results reached $r^2=0.951$ and $q^2=0.943$, which are slightly better than the present approach. Again it seems proved that for highly homogeneous series, $<V_{ee}>$ produces equivalent information to the one provided by log P.

6.3.3
Aquatic toxicity of benzene-type compounds

The following example sets have been already studied elsewhere in this book, see section 4.4.2. Here, an alternative approach to the description of their properties is presented, using $<V_{ee}>$ as a molecular descriptor instead of the full quantum similarity matrix. Even though the present new QSAR models will be simpler than the former ones, the unique descriptor $<V_{ee}>$ still provides a good description of the property. Two sets of benzene-type chemicals were examined: a set of 36 benzene derivatives and a set of 21 phenols, which present acute toxicity to animal species. The property studied is the aquatic toxicity to the fish *Poecilia reticulata* [277], measured by -log EC_{50}, the concentration necessary to reduce 50% the fish population. In contrast to the full quantum similarity matrix, $<V_{ee}>$ was not able to correlate satisfactorily these two combined sets. However, the division of them into two classes, according to structural homogeneity criteria, led to acceptable structure-activity correlations.

The first set is made up of 36 benzene derivatives, including chlorobenzenes, nitrobenzenes, toluenes and xylenes. The actual molecular set corresponds to the first 36 elements of Table 4.5, and thus they will not be reproduced here. The mathematical connection between the toxicity and the molecular descriptor is:

$$ -\log EC_{50} = -1.751\cdot 10^{-3}\langle V_{ee}\rangle - 2.667 $$

$$ n = 36 \qquad r^2 = 0.889 \qquad \sigma_N = 0.229 \qquad q^2 = 0.877 $$

As for it has been done in the previous computation cases, the plot of cross-validated $-\log EC_{50}$ versus the experimental values is presented in Figure 6.3.

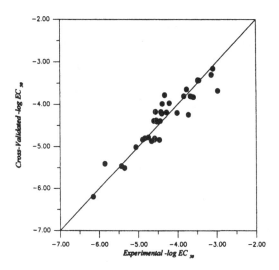

Figure 6.3. Cross-validated versus experimental –log EC_{50} for a set of 36 benzene derivatives.

The results obtained for this narcotic pollutants, r^2=0.889 and q^2=877, are excellent, regarding the fact that a unique descriptor was employed. This nice QSAR proves explicitly that $<V_{ee}>$ may discern between molecules that differ in the position of their substituents. In other words, $<V_{ee}>$ not only provides information about the molecular size of the involved molecules, but also on their internal electronic distributions. Results cannot be straightforwardly compared to those achieved with the full quantum similarity matrix, so a recalculation for this specific subset was carried out. The optimal model was obtained when using 5 PCs: r^2=0.892 and q^2=0.840. When only 1 PC was used, the results were also satisfactory: r^2=0.798 and q^2=0.771, but considerably lower than those produced by means of the $<V_{ee}>$ description.

The second example molecular set is made up of 21 phenol derivatives, where the aquatic toxicity as the preceding example [277], was associated with $<V_{ee}>$. Structures and toxicity of this molecular set are given in Table 4.5, see chapter 4, and correspond to compounds varying from phenol to quinoline. The QSAR model is defined by the following linear equation, and the quality of the model is assessed with the usual statistical parameters:

$$-\log EC_{50} = -2.712 \cdot 10^{-3} \langle V_{ee} \rangle - 2.388$$

$$n = 21 \qquad r^2 = 0.897 \qquad \sigma_N = 0.142 \qquad q^2 = 0.878$$

The leave-one-out cross-validation predictions can be plotted against the experimental values, see Figure 6.4.

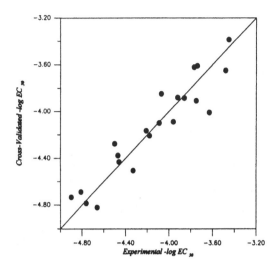

Figure 6.4. Cross-validated versus experimental –log EC_{50} for the set of 18 phenol derivatives.

As it can be easily observed, the correlation is again excellent, and it can be compared with the ones computed using other approaches and points of view. In particular, results for the phenol subset using the full similarity matrix to build QSARs were even not so good: $r^2=0.846$ and $q^2=0.746$ using 5 PCs. $<V_{ee}>$ correlation evidences again the capacity of this descriptor to discern between different structural isomers. Thus, the set of methylphenols or dimethylphenols only differ one from another in the exact position of the substituents, which also is connected to the toxic action. This isomeric influence is well characterized by the interelectronic repulsion energy. Another important remark, derived from the analysis of the narcotic pollutants, is the capacity of $<V_{ee}>$ to reproduce the information contained in hydrophobic descriptors such as log P, as it has been commented before. Even though it has not been proposed explicitly, for highly homogenous series this parameter can be also used as an alternative to log P. On the other hand, $<V_{ee}>$ is more sensitive to the homogeneity of the sets, and it is unable to deal with chemicals presenting some degree of heterogeneity in their structure. In particular, the whole set of 92 benzene derivatives, from which these two families were extracted, was impossible to attain a significant correlation with $<V_{ee}>$.

6.3.4
Activity of alkylimidazoles

Establishment of linear relationships between single descriptors and biological parameters is not always possible, and this impediment could be due to that the ordinate at the origin, b, in equation (6.9) can be considered no longer a constant.

In these cases, it is proposed to extend the linear formulation to a general polynomial formulation, in order to take into account the fluctuations in the off-diagonal MQSM, affecting the parameter b, as it can be seen in equation (6.11):

$$y_A \approx \sum_{p=0}^{k} a_p \langle V_{ee} \rangle^p + b \cdot \qquad (6.11)$$

Within this simple framework, it is sometimes possible to keep using $\langle V_{ee} \rangle$ as a simple molecular descriptor.

As an illustrative example, the biological activity of 13 alkylimidazoles was examined. Imidazoles present a high potency as inhibitors of drug oxidation [278-282], potentiators of barbiturate sleeping time in mammals [280,282,283] and insecticide synergists to houseflies [283]. In addition, the ligand interaction with cytochrome P-450 (type II) has also been analyzed. In particular, the microsomal epoxidation of aldrin in enzyme preparations from armyworm (*Prodenia eridania*) was studied. The first property, namely the armyworm gut aldrin epoxidation, is measured by $-\log I_{50}$, the inhibitor concentration necessary to produce 50% epoxidation. The second biological property consists of rat liver cytochrome p-450 (type-II) binding affinity, evaluated by the $-\log K_S$ parameter, being K_S the spectral dissociation constant. Data was taken from the work of Wilkinson et al. [284] and are reproduced in Table 6.3.

Table 6.3. Values of $\langle V_{ee} \rangle$ and biological activities for the set of 13 1-alkylimidazoles.

Alkyl group	$\langle V_{ee} \rangle$	$-\log I_{50}$	$-\log K_S$
H	238.99	2.53	3.68
CH_3	314.30	2.90	3.70
CH_2CH_3	394.39	3.27	4.68
$(CH_2)_3CH_3$	474.88	4.85	4.94
$(CH_2)_4CH_3$	558.46	5.58	5.03
$(CH_2)_5CH_3$	643.95	6.41	5.39
$(CH_2)_6CH_3$	731.99	6.87	5.47
$(CH_2)_7CH_3$	821.99	6.94	5.56
$(CH_2)_8CH_3$	914.17	7.02	5.49
$(CH_2)_9CH_3$	1008.18	7.09	5.41
$(CH_2)_{10}CH_3$	1104.06	6.99	5.46
$(CH_2)_{12}CH_3$	1300.79	6.75	5.35
$(CH_2)_{14}CH_3$	1503.63	6.08	5.24

These two properties were first analyzed as usual, connecting them linearly with $\langle V_{ee} \rangle$. The QSAR models using the common linear regressions with $\langle V_{ee} \rangle$

as a parameter yield to a very poor results: $r^2=0.557$, $q^2=0.259$ for the armyworm epoxidation and $r^2=0.477$, $q^2=0.154$ for the rat liver cytochrome binding affinity. However, the particular data distribution suggested that a quadratic relationship with $<V_{ee}>$ could improve notably the adjustment. This is due to the fact that the biological activities clearly pass through a maximum as the series is ascended, so some importance of the squared term is expected. Using a second-degree polynomial fit the QSAR models were then recalculated. Thus, the results of the quadratic regressions were:

$$-\log I_{50} = -1.039\langle V_{ee}\rangle^2 + 1.624\langle V_{ee}\rangle + 6.670$$

$n = 13$ $\qquad r^2 = 0.965$ $\qquad \sigma_N = 0.231$ $\qquad q^2 = 0.931$

$$-\log K_S = -0.399\langle V_{ee}\rangle^2 + 0.587\langle V_{ee}\rangle + 5.430$$

$n = 13$ $\qquad r^2 = 0.900$ $\qquad \sigma_N = 0.156$ $\qquad q^2 = 0.745$

In spite of the structural simplicity of the molecular series, the activity does not follow a trivial pattern, and the last additions of methylene groups do not increment the property values. However, the QSAR results for the two properties using $<V_{ee}>$ as a parameter are acceptable: $q^2=0.931$ for the armyworm epoxidation and $q^2=0.745$ for the cytochrome binding affinity. The latter activity presents two remarkable outliers: compounds **2** (15% of relative error) and **13** (14% of relative error).

Figures 6.5 and 6.6 show the cross-validation test plots for the two properties.

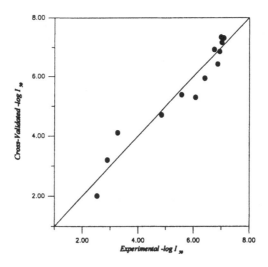

Figure 6.5. Cross-validated versus experimental $-\log I_{50}$ for a set of 1-alkylimidazoles.

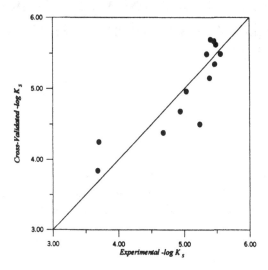

Figure 6.6. Cross-validated versus experimental –log K_S for a set of 1-alkylimidazoles.

Wilkinson et al. [284] studied the same molecular set. Curiously, these authors found a similar behavior when correlating the data. Poor correlations were attained when using a single descriptor, which in this case was the hydrophobicity descriptor π [285]. Correlation was considerably improved when including a π^2 term. The quality models for the *in vitro* aldrin epoxidation from armyworm were $r^2=0.600$ with the linear equation, and $r^2=0.966$ with the quadratic equation. For the rat liver cytochrome binding affinity, the results were $r^2=0.520$ using the linear equation and $r^2=0.924$ when adding the quadratic term. These authors did not perform cross-validation, and q^2 coefficient was not reported. The use of a hydrophobicity descriptor was due to the fact that lipophilic character determines the ability of compounds to interact with microsomal drug-metabolizing enzymes [286]. Different studies assessed that a good correlation exist between the substrate binding (K_m) and the hydrophobic parameter log P [287,288]. In this example, one retrieves the same behavior as before: $<V_{ee}>$ is able, for homogeneous molecular series, to produce an analogous information as hydrophobic descriptors, with the obvious advantage that the electron-electron repulsion energy can be exactly computed for any molecule with available quantum-chemical software.

7 Quantum Similarity extensions to non-molecular systems: Nuclear Quantum Similarity

So far, Quantum Similarity has only been applied to a chemical environment. However, its theoretical basis is flexible enough to allow a satisfactory extension to other quantum objects sets. In this chapter, application of Quantum Similarity to atomic nuclei and its use in quantitative structure-property relationships are discussed.

7.1
Generality of Quantum Similarity for quantum systems

In the initial chapters, the words *quantum object* have been deliberately used when referring to the type of objects that can be analyzed within the Quantum Similarity framework. After this, however, the discussed applications have been restricted to the field of chemistry, and therefore *quantum object* was looking equivalent to *molecule*. Nevertheless, any type of microscopic system can be subdued to a Quantum Similarity analysis, provided it is feasible to be described in a quantum-mechanical way, that is, by a wavefunction, and in extension, by a density function. The exact form of the density function depends on the fields that act over the system, which constitute the different terms of the Schrödinger equation: kinetic energy, electrostatic potential, strong force (for nucleons), etc. Thus, the application universe of quantum similarity can be extended to atoms, nuclei, molecular nuclei or infinite nuclear matter. In this book, an extension of Quantum Similarity to atomic nuclei is discussed, giving some application examples.

7.2
Nuclear Quantum Similarity

In this section, an extension of Quantum Similarity to nuclear physics is discussed. A first attempt to extend Quantum Similarity to nuclear systems was proposed by Robert and Carbó-Dorca [289], and most of the concepts discussed in that paper are reproduced here. The methodology employed is exactly the same as in the molecular case, but the form and the way of deriving the density functions is different. In addition, the theoretical framework for nuclear QSAR is also identical to the chemical case. As it will be seen, interesting results are achieved with this approach.

For the sake of simplicity, only overlap-like measures have been used:

$$Z_{AB} = \int \rho_A(\mathbf{r}) \rho_B(\mathbf{r}) \, d\mathbf{r} \tag{7.1}$$

By a logical extension, these measures will be referred to as *nuclear quantum similarity measures* (NQSM).

7.2.1
The nuclear density functions: the Skyrme-Hartree-Fock model

The lack of information of the exact form of the nucleon-nucleon potential has lead to elaborate sophisticated models to describe the nuclear ground state [290]. One of the existing approaches consists of the use of Hartree-Fock calculations with effective interactions, which are parameterized using experimental data of relevant nuclear magnitudes, such as the nuclear matter incompressibility or Fermi momentum [291]. This methodology is less fundamental than using a realistic force, but it has produced accurate results. From all the proposed interactions, the one proposed by T.H.R Skyrme in 1959 [292] today still prevails due to the model ability to provide a good description of the ground state of spherical nuclei. This approach, namely Hartree-Fock calculations with an effective Skyrme force [293], has been used to obtain the nuclear density functions, which serve as a basis for the quantum similarity calculations in this work.

No many details will be given about this theoretical nuclear model, as this task is far from the general purpose of this study. In general lines, the Skyrme interaction is an effective force that depends on the density, with a short-range two-body term and a three-body term. The Skyrme force depends on a small number of parameters, and since its appearance, several parameterizations have been proposed in order to get the best possible description of the nuclear ground state. From all of them, it has been chosen the SkM* parameterization, proposed by Bartel et al. [294].

The simplicity of this ansatz makes possible to express the Hamiltonian as a function of few densities, such as the kinetic energy density, the spin density, the divergence of the spin density or the nuclear matter density, which constitutes our objective. All the nuclei are treated as if they were spherical, by introducing state occupation probabilities, which are computed by means of a simple pairing scheme [295]. This yields the so-called HF+BCS equations. The mean field localizes the nucleus and breaks translational invariance, and in consequence the nuclear ground state is not a state with zero total momentum, but a state in which the whole nucleus oscillates in the mean field [296]. A state with good zero total momentum out of the given mean field can be simply projected by introducing a center-of-mass correction as a modification of the nucleon mass.

The Hartree-Fock equations have been solved by the usual self-consistent method. The number of cycles has been chosen in such a way that four-point energy self-consistency is achieved. The solutions of the analytical Woods-Saxon potential [297] have been used as trial wavefunctions. Proton and neutron masses have been taken as equal. The solution of the density function used a radial grid of 150 points; with an interpoint distance of 0.1 fm. Figure 7.1 shows some of the nuclear matter densities.

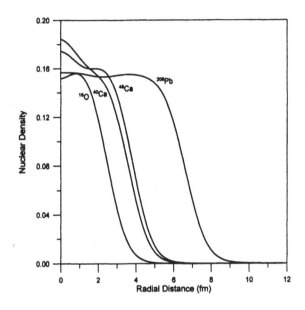

Fig. 7.1. Skyrme-Hartree-Fock nuclear density functions for the ^{16}O, ^{40}Ca, ^{48}Ca and ^{208}Pb.

By analogy, the quantum similarity measures related to comparisons between atomic nuclei will be referred to as *nuclear quantum similarity measures* (NQSM). In this work, only overlap-like NQSM will be used. Note that the spherical symmetry for nuclei transforms the NQSM into a simplified form:

$$Z_{AB} = \int \rho_A(\mathbf{r}) \rho_B(\mathbf{r}) d\mathbf{r} = \int \rho_A(r) \rho_B(r) 4\pi r^2 \, dr \qquad (7.2)$$

The NQSM integrals have been computed numerically using the second-order Simpson rule.

7.3
Structure-property relationships in nuclei

The major part of the physical properties of nuclei is strongly related to its structure, and simple models are able to describe them with a high accuracy. Nevertheless, there are certain properties in nuclear physics that are not so easy to derive from theoretical state approaches, and sophisticated semiempirical models are then constructed to give an answer to this problem. Binding energy per nucleon and mass excess are two examples of this kind of properties.

In this section, Quantum Similarity is presented as a possible alternative to the usual semiempirical models. The way of extracting information from NQSM is identical to the chemical procedure as outlined in chapter 4. The full quantum similarity matrix is dealt with classical scaling, but no selection of variables has been performed here. The reason is that nuclear symmetry makes the information to be focused in few PCs, and more than 90% of variance is accounted for the first three PCs. Then, the PCs are used as parameters in a multilinear regression. The robustness of the models are again tested with the presented statistical methods: cross-validation and randomization, see sections 3.5 and 3.6 for definition.

7.3.1
The nuclear data set

For all the studied properties, a medium-size nuclear set has been employed. Such data set is made up of 73 stable nuclei found in the range of light and semiheavy regions within the nuclear periodic table: concretely from the ^{16}O up to the ^{208}Pb. This range is the optimal one for the application of the Skyrme-Hartree-Fock model, the one used to describe the nuclear ground state in this study, and eventually, to construct the nuclear density functions used as a source of the NQSM. As it has been stated, this theoretical approach describes poorly the heavy (deformed) nuclei, and those that are far from the stability line. The selection of nuclei includes *isotopes* of a same element, such as ^{24}Mg and ^{25}Mg; or ^{206}Pb, ^{207}Pb and ^{208}Pb, among others. It also includes nuclei with identical massic number but a different number of protons and neutrons, the so-called *isobar nuclei*, such as ^{48}Ca and ^{48}Ti.

7.3.2
The binding energy per nucleon

Binding energy per nucleon is a roughly constant property (from −7.8 to −8.8 MeV per nucleon, except for very light nuclei), with a smooth peak near A=56. This value indicates where the nuclei are most tightly bound, and therefore, the binding energy per nucleon can be considered, in some sense, a measure of the stability of a nuclear configuration. Experimental data have been taken from the

Audi and Wapstra's compilation of experimental works by several authors [298]. Table 7.1 gathers these values.

The relationship between the binding energy per nucleon and the number of nucleons is not simple, as can be deduced from Figure 7.2. In this case, this property cannot be expressed by means of a linear or 2-degree polynomial regression using the massic number, and in fact, the existing models are more complex [299,300].

Table 7.1. Binding energy per nucleon (MeV per nucleon) for the 73 studied nuclei.

Nucleus	Observed E/A (MeV)	Nucleus	Observed E/A (MeV)	Nucleus	Observed E/A (MeV)	Nucleus	Observed E/A (MeV)
^{16}O	-7.9762	^{48}Ti	-8.7229	^{89}Y	-8.7139	^{169}Tm	-8.1145
^{18}O	-7.7671	^{51}V	-8.7420	^{90}Zr	-8.7099	^{175}Lu	-8.0692
^{19}F	-7.7790	^{52}Cr	-8.7759	^{93}Nb	-8.6641	^{181}Ta	-8.0234
^{20}Ne	-8.0322	^{55}Mn	-8.7649	^{103}Rh	-8.5841	^{185}Re	-7.9910
^{24}Ne	-7.9932	^{56}Fe	-8.7902	^{107}Ag	-8.5539	^{187}Re	-7.9780
^{23}Na	-8.1115	^{59}Co	-8.7679	^{109}Ag	-8.5479	^{191}Ir	-7.9481
^{24}Mg	-8.2607	^{58}Ni	-8.7320	^{115}In	-8.5166	^{193}Ir	-7.9381
^{25}Mg	-8.2235	^{60}Ni	-8.7807	^{121}Sb	-8.4820	^{194}Pt	-7.9360
^{27}Al	-8.3316	^{63}Cu	-8.7521	^{123}Sb	-8.4723	^{195}Pt	-7.9267
^{28}Si	-8.4477	^{69}Ga	-8.7245	^{127}I	-8.4455	^{196}Pt	-7.9266
^{31}P	-8.4812	^{71}Ga	-8.7176	^{133}Cs	-8.4100	^{197}Au	-7.9157
^{32}S	-8.4931	^{75}As	-8.7009	^{138}Ba	-8.3935	^{203}Tl	-7.8861
^{35}Cl	-8.5203	^{78}Se	-8.7178	^{139}La	-8.3781	^{205}Tl	-7.8785
^{37}Cl	-8.5703	^{80}Se	-8.7108	^{140}Ce	-8.3764	^{206}Pb	-7.8754
^{40}Ar	-8.5953	^{79}Br	-8.6876	^{141}Pr	-8.3541	^{207}Pb	-7.8699
^{39}K	-8.5570	^{81}Br	-8.6959	^{151}Eu	-8.2394	^{208}Pb	-7.8675
^{40}Ca	-8.5513	^{85}Rb	-8.6974	^{153}Eu	-8.2288		
^{48}Ca	-8.6665	^{87}Rb	-8.7110	^{159}Tb	-8.1889		
^{45}Sc	-8.6189	^{88}Sr	-8.7326	^{165}Ho	-8.1470		

There exist different theoretical approaches to estimate the binding energy per nucleon, everyone yielding excellent results. One of the most employed is the expression proposed by Weizsäcker [299] and Bethe and Bacher [300], which describes the binding energy with a semiempirical formula, sum of different contributions:

$$E = a_V A - a_S A^{2/3} - a_I \frac{(Z-N)^2}{A} - a_C \frac{Z^2}{A^{1/3}} + \delta \qquad (7.3)$$

where δ is a parameter which can take on the following values:

$$\delta = +\frac{a_p}{A^{1/2}} \quad \text{for Z, N even}$$

$$\delta = -\frac{a_p}{A^{1/2}} \quad \text{for Z, N odd} \tag{7.4}$$

$$\delta = 0 \qquad \text{for any other case}$$

One of the proposed parameterizations is: a_V=−14.1 MeV, a_s=−13.0 MeV, a_C=−0.595 MeV, a_I=−76 MeV, a_p=−12 MeV.

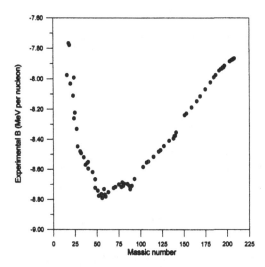

Fig 7.2. Representation of the binding energy per nucleon (MeV/nucleon) versus the massic number for the 73 studied nuclei.

This equation is easily interpretable as a series of contributions to the energy. Thus, the first term represents the volume contribution, which is directly related to the massic number. This relationship is easily explained due to the tendency of the nuclear radius: $R=R_0A^{1/3}$. Volume is proportional to R^3, and therefore, to A. The second term can be considered as a surface contribution, which has the exponent 2/3. The third term is a symmetry term, which encourages the equality between the number of protons and neutrons. This symmetry term arises due to a calculation with the Fermi gas, which shows that, fixing the massic number, the binding energy presents a maximum at $Z=N$ [301]. The following summand represents the contribution associated to the Coulomb repulsion, which obviously only affects the protons of the nuclei. This term is simply the electrostatic energy of a homogeneously charged sphere with radius $A^{1/3}$, except for numerical factors included in the parameter a_C. Finally, the parameter δ quantifies the tendency of neutrons and protons to group in couples. This five-parameter formula describes the values of E with an accuracy of 1%.

Here, NQSM will be used to construct an alternative estimation model. The studied property is the binding energy per nucleon, that is, E/A. Robert and Carbó-Dorca [302] recently reported the results of this study. Classical scaling transformation of the overlap quantum similarity matrix reveals that the variance is strongly focused, 98%, in the first two PCs, hence no variable selection was employed in order to avoid an undesirable background noise parameterization. The QSAR models are shown in Table 7.2.

Table 7.2. QSAR models for the 73 nuclei.

Number of PCs	r^2	q^2	σ_N
2	0.891	0.878	0.107
3	0.949	0.940	0.073
4	0.957	0.948	0.068

As it can be seen, excellent results are obtained with only three parameters, as indicated by the statistical descriptors: r^2=0.949, q^2=0.940. The equation for the optimal model is:

$$y = -0.105x_1 - 0.499x_2 - 0.450x_3 - 8.376 \qquad (7.5)$$

Figure 7.3 shows the graphical representation of the cross-validated binding energies versus the experimental ones.

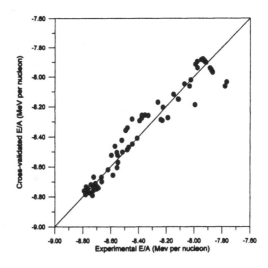

Fig 7.3. Cross-validated versus experimental binding energies per nucleon (MeV per nucleon).

No errors higher than 4% are accounted with the optimal model, slightly worse than those derived from the semiempirical formula exposed previously. In any case, from figure 7.3 it is clearly observed that the poorest adjusted points correspond to those nuclei located at the extremes of the studied set. That is: the lightest and heaviest nuclei. It is precisely in this region where the model used to build the densities, namely the Skyrme-Hartree-Fock model, cannot be applied accurately. On the other hand, medium size nuclei are reasonably well estimated, taking into account that they are described with an expression with less parameters than the semiempirical model.

This model has been validated with the randomization test, in order to detect possible chance correlations or overparametrization. The results are given in figure 7.4. As can be observed, the separation between the predictive models with the corrected responses and those with randomized ones is very clear. It can then be concluded that the achieved structure-property correlations have a real basis.

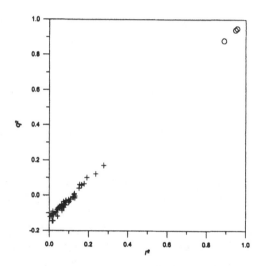

Fig. 7.4. Randomization test for the 73 nuclei. The corrected property has been marked with a circle, and the randomized vectors with crosses.

7.3.3
The mass excess

The mass of the nuclear species, expressed in atomic mass units (amu) is close to the number of nucleons, but it does not coincide exactly. A well-studied property in nuclear physics is the so-called *mass excess* (Δ), defined as:

$$\Delta = M \left(\text{in a.m.u.}\right) - A \tag{7.6}$$

Being 1 amu the atomic mass of the carbon-12 divided by 12. This property is measured in MeV. Considering the studied range, the mass excess is comprised between 1 MeV for light nuclei, and 90 MeV for heavy atoms. The dependence of this property to the massic number is not trivial, as can be deduced from figure 7.5.

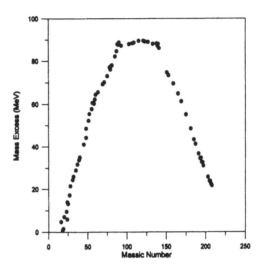

Fig 7.5. Representation of the mass excess (MeV) versus the massic number for the 73 nuclei

The computational details for the Quantum Similarity calculations are equal to the previous application example: the original overlap NQSM has been dealt with the classical scaling analysis, without any transformation into MQSI. Again the selection of variables is not justified, since the first three PCs contain 99.7% of the explained variance. The results achieved with the Quantum Similarity techniques are relatively accurate, as Table 7.3 shows. With few parameters, the models are able to estimate in a very good approximation the mass excesses for the analyzed nuclei. Following the criteria exposed in the theoretical chapters, the 4-parameter model has been taken as the optimal one in terms of balance between the number of parameters and the capacity of description. The equation for the optimal model is:

$$y = 4.339x_1 - 51.484x_2 + 26.810x_3 - 32.135x_4 - 53.075 \qquad (7.7)$$

Table 7.3. Statistical results of the predictive models

Number of PCs	r^2	q^2	σ_N
2	0.962	0.958	5.460
3	0.990	0.988	2.585
4	0.996	0.995	1.841
5	0.996	0.995	1.784

In order to illustrate better the adjustments of the models, figure 7.6 shows the cross-validated mass excesses versus the experimental ones for the 73 nuclei, using equation (7.7).

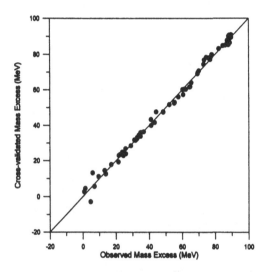

Fig. 7.6. Cross-validated versus experimental mass excesses for the 73 nuclei.

The verification that there have not been obtained spurious correlations has been carried out by means of the randomization test. Figure 7.7 shows the values of the coefficients r^2 and q^2 for the property vector correctly and randomly ordered. The discrimination between both situations is very clear, and leads to a rejection of the chance correlation hypothesis.

The precision in the predictions of the mass excess seems very high, but it needs to be compared to that achieved by existing theoretical models. Haustein [303] compiled the results for the main predictive models. Ten models were discussed in that paper. Two of the models did not encompass the nuclear range studied here, and they will not be commented.

The main feature of all of these models is the huge accuracy and the large range they can be applied to. On the other hand, the price to pay is the high number of parameters employed. The results obtained with the Quantum Similarity approach are less ambitious: they deal with a restricted region of the nuclear periodic table and the generated predictions are not so good, but the models are considerably simpler.

Table 7.4 shows the general characteristics of the theoretical models: number of parameters, training and test set, standard deviation and root mean squared (rms) deviation.

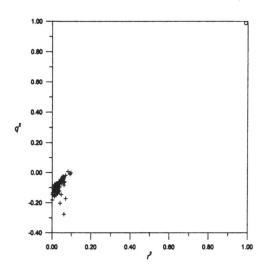

Fig. 7.7. Randomization test for the 73 nuclei. Correct properties have been marked with a circle (o) and randomized ones with a cross (+).

In the following, the different existing models for the prediction of the mass excess will be briefly described. Möller, Myers, Swiatecki and Treiner proposed a nuclear mass formula that used a finite range droplet model and a Folded-Yukawa single-body potential. With relatively few parameters, 29, described a large range of nuclei. The model gives excellent results for those species near the stability line, and worse for those far from it. Möller and Nix proposed a unified model with macroscopic and microscopic elements. It also employed few descriptors (26), and the results follow similar trends. The extrapolation with these models needs to be done, therefore, with caution. Comay, Kelson and Zidon studied a wider range of nuclei, 6537, achieving very good predictions: rms=0.431 MeV.

The rest of the models utilize many more parameters, sacrificing the simplicity of the regression in return for a higher accuracy. Thus, Satpathy and Nayak calculated a mass formula within the infinite nuclear mass model framework, with 238 parameters. Excellent adjustments were achieved, quantified by the factor rms=0.456 MeV. Tachibana, Uno, M. Yamada and S. Yamada proposed an empirical formula that included the effects of proton-neutron interaction. The characteristics of their model, in relation to the number of descriptors and the precision in the results, are close to the last one. This model was extrapolated to a huge nuclear set: 7204 nuclear species. Jänecke and Masson used the mass relations of Garvey-Kelson to generate a model with 928 parameters. The results, taking into account the large set of analyzed nuclei, are clearly the best ones, with rms=0.343 MeV. The model of Masson and Jänecke, based on an equation that includes high-order isospin contributions, achieves similar results than the previous models, with almost half the descriptors, 471.

An interesting detail to remark is that the models of Comay et al., Jänecke et al. and Masson et al., obtain relatively few isotopes with a high deviation. This fact yields to small values of the rms coefficient. Alternatively, both two models of Möller et al., exhibit many more isotopes with diverse deviations, which explains the high value of the rms coefficient.

Table 7.4. General features of the different theoretical models for the prediction of the mass excess.

Model	Number of parameters	Training set	rms deviation (keV)	Mean deviation[a] (keV)	Test set[b]
NQSM[c]	4	73	15733	0[d]	73
Möller, Myers, Swiatecki, Treiner	29	1593	777	14	4635
Möller, Nix	26	1593	849	13	4635
Comay, Kelson, Zidon	—	1541	431	10	6537
Satpathy, Nayak	238	1593	456	1	3481
Tachibana, Uno, Yamada, Yamada	281	1657	538	22	7204
Jänecke, Masson	928	1535	343	16	5860
Masson, Jänecke	471	1504	346	14	4383

[a] *Based on the convention: deviation = calculated mass – experimental mass*
[b] *Includes the training set*
[c] *Deviations obtained from the adjustment*
[d] *By construction*

The previous results point out that the philosophy of these approaches is quite different from Quantum Similarity treatment. The previously discussed equations do not pretend to provide a physical sense to anyone of their parameters, but to obtain regressions able to get perfect extrapolations, regardless of the number of descriptors. It must be emphasized that the proposed methodology is strongly sensitive to the theoretical model which generates the nuclear density functions, but at the same time remains opened to model improvements. The attached results, therefore, are not definitive, but can presumably be improved with more sophisticated tools.

Despite that quantum similarity results are poorer than those attained by other models, various factors must be considered to get a complete assessment of the methodology: first, the number of descriptors used in the rest of models is larger than in the NQSM approach –up to two magnitude orders–, and subsequently, they cannot be considered bad results due to the simplicity of the regression. Second, the proposed equations here do not correspond to closed and invariable models. The major goal of the method is establishing a general methodology to extract information from the density functions of the quantum objects and in this way following quantum theoretical procedures. Thus, an improvement of the theoretical models which describe the nuclear interactions will lead to an increase in the accuracy of the nuclear density functions, and subsequently, it will necessarily augment the goodness of the adjustments and predictions of the nuclear properties.

7.4
Limitations of the approach

Finally, it is important to remark the limitations of the methodology. It has been proved that Quantum Similarity is a useful tool for extracting information from the nuclear densities by means of comparative measures between these functions. The main drawback of this approach is the little development of the techniques to build the nuclear densities, due to the poor knowledge of the nucleon-nucleon interaction inside the atomic nucleus. In any case, the results provided as an example are not definitive, and analogous calculations could be performed using density functions derived from more accurate theoretical models. Within the properties studied, it has been evidenced the incapacity of the Skyrme interaction to correctly describe light nuclei, where the clustering effects inside the nucleus become important, or heavy nuclei, where the presence of deformations suggests the use of theoretical models capable to describe nuclear phenomena as a collective. In addition, there have been initially discarded those nuclear species far from the stability line, as well as those very heavy, such as the uranides and transuranides. Together with the intrinsic limitations of the model, it must also be remarked the approximations made. Thus, considering the proton and neutron mass differences or a denser grid, could improve the accuracy of the derived results.

References

Chapter 1

1. Carbó R, Leyda L, Arnau M (1980) How similar is a molecule to another? An electron density measure of similarity between two electronic structures. Int J Quantum Chem 17:1185-1189
2. Carbó R, Domingo L (1987) LCAO-MO similarity measures and taxonomy. Int J Quantum Chem 23:517-545
3. Carbó R, Calabuig B (1989) MOLSIMIL-88: Molecular similarity calculations using a CNDO-like approximation. Comput Phys Commun 55:117-126
4. Carbó R, Calabuig B (1992) Molecular quantum similarity measures and n-dimensional representation of quantum objects. I. Theoretical foundations. Int J Quantum Chem 42:1681-1693
5. Carbó R, Calabuig B (1992) Molecular quantum similarity measures and n-dimensional representation of quantum objects. II. Practical applications. Int J Quantum Chem 42:1695-1709
6. Carbó R, Calabuig B, Besalú E, Vera L (1994) Molecular quantum similarity: theoretical framework, ordering principle and visualization techniques. Adv Quantum Chem 25:255-313
7. Besalú E, Carbó R, Mestres J, Solà M (1995) Foundations and recent developments on molecular quantum similarity. Top Curr Chem 173:31-62
8. Carbó-Dorca R, Besalú E (1998) A general survey of molecular quantum similarity. J Mol Struct (Theochem) 451:11-23
9. Carbó-Dorca R, Amat L, Besalú E, Lobato M (1998) Quantum similarity. In: Carbó-Dorca R, Mezey PG (eds) Advances in molecular similarity. JAI Press, Greenwich, Vol 2, pp 1-42
10. Carbó-Dorca R, Amat L, Besalú E, Gironés X, Robert D (1999) Quantum mechanical origin of QSAR: theory and applications. J Mol Struct (Theochem) in press
11. Cioslowski J, Fleischmann ED (1991) Assessing molecular similarity from results of ab initio electronic structure calculations. J Am Chem Soc 113:64-67
12. Cioslowski J, Challacombe M (1991) Maximum similarity orbitals for analysis of the electronic excited states. Int J Quantum Chem 25:81-93
13. Ortiz JV, Cioslowski J (1991) Molecular similarity indices in electron propagator theory. Chem Phys Lett 185:270-275
14. Cioslowski J (1992) Differential density matrix overlap: an index for assessment of electron correlation in atoms and molecules. Theor Chim Acta 81:319-327
15. Cooper DL, Allan NL (1989) A novel approach to molecular similarity. J Comput-Aided Mol Design 3:253-259
16. Cooper DL, Allan NL (1992) Molecular dissimilarity: a momentum-space criterion. J Am Chem Soc 114:4773-4776
17. Allan NL, Cooper DL (1992) A momentum space approach to molecular similarity. J Chem Inf Comput Sci 32:587-590
18. Allan NL, Cooper DL (1995) Momentum-space electron densities and quantum molecular similarity. Top Curr Chem 173: 85-111

19. Cooper DL, Allan NL (1995) Molecular similarity and momentum space. In: Carbó R (ed) Molecular similarity and reactivity: from quantum-chemical to phenomenological approaches. Kluwer, Amsterdam, pp 31-55

20. Measures PT, Allan NL, Cooper DL (1996) Momentum-space similarity: some recent applications. In: Carbó R, Mezey PG (eds) Advances in molecular similarity. JAI Press, Greenwich, Vol 2, pp 61-87

21. Allan NL, Cooper DL (1998) Quantum molecular similarity via momentum-space indices. J Math Chem 23:51-60

22. Bowen-Jenkins PE, Richards WG (1985) Ab initio computation of molecular similarity. J Phys Chem 89:2195-2197

23. Hodgkin EE, Richards WG (1987) Molecular similarity based on electrostatic potential and electric field. Int J Quantum Chem 14:105-110

24. Meyer AM, Richards WG (1991) Similarity of molecular shape. J Comput-Aided Mol Design 5:426-439

25. Richards WG (1995) The dominant role of shape similarity and dissimilarity in QSAR. In: Sanz F, Manaut F (eds) QSAR and molecular modelling: concepts, computational tools and biological applications. Prous Science, Barcelona, pp 364-373

26. Herndon, WC (1988) Graph codes and a definition of structural similarity. Comput Math Applic 15:303-309

27. Mezey PG (1993) Shape in Chemistry: an introduction to molecular shape and topology. VCH, New York

28. Mezey PG (1988) Shape group studies of molecular similarity: shape groups and shape graphs of molecular contour surfaces. J Math Chem 2:299-323

29. Arteca GA, Mezey PG (1989) Molecular similarity and molecular shape changes along reaction paths: a topological analysis and consequences on the Hammond postulate. J Phys Chem 93:4746-4751

30. Mezey PG (1991) The degree of similarity of three-dimensional bodies: applications to molecular shapes. In: Mezey PG (ed) Mathematical Modeling in Chemistry. VCH, New York, pp 39-49

31. Mezey PG (1992) Shape-similarity measures for molecular bodies: a 3D topological approach to QShAR. J Chem Inf Comput Sci 32:650-656

32. Ponec R (1987) Topological aspects of chemical reactivity. On the similarity of molecular structures. Collect Czech Chem Commun 52:555-561

33. Ponec R, Strnad M (1990) Similarity approach to chemical reactivity. Specificity of multibond reactions. Collect Czech Chem Commun 55:2583-2589

34. Ponec R, Strnad M (1991) Topological aspects of chemical reactivity. Evans/Dewar principle in terms of molecular similarity approach. J Phys Org Chem 4:701-705

35. Ponec R, Strnad M (1992) Electron correlation in pericyclic reactivity: a similarity approach. Int J Quantum Chem 42:501-508

36. Ponec R, Strnad M (1993) Position invariant index for assessment of molecular similarity. Croat Chem Acta 66:123-127

37. Mezey PG, Ponec R, Amat L, Carbó-Dorca R (1999) Quantum similarity approach to the characterization of molecular chirality. Enantiomeres 4:371-378

38. Constans P, Amat L, Carbó-Dorca R (1997) Toward a global maximization of the molecular similarity function: superposition of two molecules. J Comput Chem 18:826-846

39. Solà M, Mestres J, Carbó R, Duran M (1994) Use of ab initio quantum similarity measures as an interpretative tool for the study of chemical reactions. J Am Chem Soc 116:5909-5915

40. Solà M, Mestres J, Carbó R, Duran M (1996) A comparative analysis by means of quantum molecular similarity measures of density distributions derived from conventional ab initio and density functional methods. J Chem Phys 104:636-647

41. Forés M, Duran M, Solà M (1997) A procedure for assessing the quality of a given basis set based on quantum molecular similarity measures. Theor Mol Mod Electr Conf 1:50-56

42. Carbó R, Besalú E, Amat L, Fradera X (1995) Quantum molecular similarity measures (QMSM) as a natural way leading towards a theoretical foundation of quantitative structure-properties relationship. J Math Chem 18:237-246

43. Fradera X, Amat L, Besalú E, Carbó-Dorca R (1997) Application of molecular quantum similarity to QSAR. Quant Struct-Act Relat 16:25-32

44. Lobato M, Amat L, Besalú E, Carbó-Dorca R (1997) Structure-activity relationships of a steroid family using quantum similarity measures and topological quantum similarity indices. Quant Struct-Act Relat 16:465-472

45. Amat L, Robert D, Besalu E, Carbó-Dorca R (1998) Molecular quantum similarity measures tuned QSAR: An antitumoral family validation study. J Chem Inf Comput Sci 38:624-631

46. Amat L, Carbó-Dorca R, Ponec R (1998) Molecular quantum similarity measures as an alternative to log P values in QSAR studies. J Comput Chem 19:1575-1583

47. Robert D, Amat L, Carbó-Dorca R (1999) Three-dimensional quantitative structure-activity relationships from tuned molecular quantum similarity measures: Prediction of the corticosteroid binding globulin binding affinity for a steroid family. J Chem Inf Comput Sci 39:333-344

48. Ponec R, Amat L, Carbó-Dorca R (1999) Molecular basis of quantitative structure-properties relationship (QSPR): A quantum similarity approach. J Comput-Aided Mol Design 13:259-270

49. Ponec R, Amat L, Carbó-Dorca R (1999) Quantum similarity approach to LFER: Substituent and solvent effects on the acidities of carboxylic acids. J Phys Org Chem 12:447-454

50. Amat L, Carbó-Dorca R, Ponec R (1999) Simple linear models based on quantum similarity measures. J Med Chem 42:5169-5180

51. Robert D, Gironés X, Carbó-Dorca R (1999) Facet diagrams for quantum similarity data. J Comput-Aided Mol Design 13:597-610

52. Robert D, Carbó-Dorca R (1999) Aromatic compounds aquatic toxicity QSAR using molecular quantum similarity measures. SAR QSAR Environ Res 10:401-422

53. Gironés X, Amat L, Carbó-Dorca R (1999) Using molecular quantum similarity measures as descriptors in quantitative structure-toxicity relationships. SAR QSAR Environ Res, in press

54. Cioslowski J, Stefanov BB, Constans P (1996) Efficient algorithm for quantitative assessment of similarities among atoms in molecules. J Comput Chem 17:1352-1358

55. Solà M, Mestres J, Oliva JM, Duran M, Carbó-Dorca M (1996) The use of ab initio quantum molecular self-similarity measures to analyze electronic charge density distributions. Int J Quantum Chem 58:361-372

56. Robert D, Carbó-Dorca R (1998) On the extension of quantum similarity to atomic nuclei: nuclear quantum similarity. J Math Chem 23:327-351

57. Robert D, Carbó-Dorca R (1999) Structure-property relationships in nuclei. Prediction of the binding energy per nucleon using a quantum similarity approach. Nuovo Cimento A111:1311-1321

58. Fradera X, Duran M, Mestres J (1998) Second-order quantum similarity measures from intracule and extracule densities. Theor Chem Acc 99:44-52

112

59. Quoted in: Borman S (1990) New QSAR techniques eyed for environmental assessments. Chem Eng News 68:20-23
60. Crum-Brown A, Fraser TR (1868) On the connection between chemical constitution and physiological action. Part 1. On the physiological action of the ammonium bases, derived from Strychia, Brucia, Thebaia, Codeia, Morphia and Nicotia. Trans Royal Soc Edinburgh 25:257-274
61. Richet C (1893) C R Seánces Soc Biol 9:775
62. Meyer H (1899) Theorie der Alkoholnarkose, welche Eigenschaft die Anästhetica bedingt ihre narkotische Wirkung Arch Exp Pathol Pharmakol 42:109-118
63. Overton E (1901) Studien über die Narkose. Gustav Fischer, Jena
64. Ferguson J (1939) The use of chemical potentials as indicators of toxicity. Proc Royal Soc London B127:387
65. Hammett LP (1937) The Effect of Structures upon the Reactions of Organic Compounds. Benzene Derivatives. J Am Chem Soc 59:96
66. Hammett LP (1940) Physical Organic Chemistry. McGraw-Hill, New York
67. Taft RW (1952) J Am Chem Soc 74:3120-3128
68. Free SM, Wilson JW (1964) A mathematical contribution to structure-activity studies. J Med Chem 7:395-399
69. Hansch C, Fujita T (1964) ρ-σ-π analysis. A method for the correlation of biological activity and chemical structure. J Am Chem Soc 86:1616-1626
70. Kubinyi H (1993) QSAR: Hansch analysis and related approaches. VCH, Weinheim
71. Wiener H (1947) Structural determination of paraffin boiling points. J Chem Phys 69:17-20
72. Kier LB, Hall LH, Murray WJ, Randic M (1975) Molecular connectivity. I: Relationship to nonspecific local anaesthesia. J Pharm Sci 64:1971-1974
73. Randic M (1975) On characterization of molecular branching. J Am Chem Soc 97:6609-6615
74. Kier LB, Hall LH (1976) Molecular connectivity in chemistry and drug research. Academic Press, New York
75. Kier LB, Hall LH (1986) Molecular Connectivity in Structure-Activity Analysis. Research Studies Press, Letchwork
76. Cramer III RD, Paterson DE, Bunce JD (1988) Comparative molecular field analysis (CoMFA). 1. Effect of shape on binding of steroids to carrier proteins. J Am Chem Soc 110:5959-5967
77. Klebe G, Abraham U, Mietzner T (1994) Molecular similarity indices in a comparative analysis (CoMSIA) of drug molecules to correlate and predict their biological activity. J Med Chem 37:4130-4146
78. Silverman BD, Platt DE (1996) Comparative molecular moment analysis (CoMMA): 3D-QSAR without molecular superposition. J Med Chem 39:2129-2140
79. Jain AN, Koile K, Chapman D (1994) Compass: predicting biological activities from molecular surface properties. Performance comparisons on a steroid benchmark. J Med Chem 37:2315-2327
80. Bravi G, Gancia E, Mascagni P, Pegna M, Todeschini R, Zaliani A (1997) MS-WHIM, new 3D theoretical descriptors derived from molecular surface properties: A comparative 3D QSAR study in a series of steroids. J Comput-Aided Mol Design 11:79-92
81. Kellogg GE, Kier LB, Gaillard P, Hall LH (1996) E-state fields: Applications to 3D QSAR. J Comput-Aided Mol Design 10:513-520
82. Luque FJ, Sanz F, Illas F, Pouplana R, Smeyers YG (1988) Relationships between the activity of some H2-receptor agonists of histamine and their ab initio molecular electrostatic potential (MEP) and electron density comparison coefficients. Eur J Med Chem 23:7-10

83. Richard AM (1991) Quantitative comparison of molecular electrostatic potentials for structure activity studies. J Comput Chem 12:959-969
84. Rum G, Herndon WC (1991) Molecular similarity concepts. 5. Analysis of steroid-protein binding constants. J Am Chem Soc 113:9055-9060
85. Good AC, So SS, Richards WG (1993) Structure-activity relationships from molecular similarity matrices. J Med Chem 36:433-438
86. Good AC, Peterson SJ, Richards WG (1993) QSAR's from similarity matrices. Technique validation and application in the comparison of different similarity evaluation methods. J Med Chem 36:2929-2937
87. Good AC, Richards WG (1996) The extension and application of molecular similarity to drug design. Drug Information Journal 30:371-388
88. Carbó R, Calabuig B (1992) Quantum similarity measures, molecular cloud description and structure-properties relationships. J Chem Inf Comput Sci 32:600
89. Besalú E, Amat L, Fradera X, Carbó R (1995) An application of the molecular quantum similarity: Ordering of some properties of the hexanes. In: Sanz F, Manaut M (eds) QSAR and molecular modelling: concepts, computational tools and biological applications. Prous Science: Barcelona

Chapter 2

89. Von Neumann J (1955) Mathematical Foundations of Quantum Mechanics. Princeton University Press, Princeton
90. Bohm D (1989) Quantum Theory. Dover Publications, New York
91. Goldstein S (1988) Quantum Theory without Observers –Part One. Physics Today March: 42-46, Part Two, April:38-42
92. Bell JS (1993) Speakable and Unspeakable in Quantum Mechanics- Cambridge University Press, Cambridge
93. Vinogradov IM (ed) (1987) Encyclopaedia of Mathematics. Vol. 8. Reidel-Kluwer, Dordrecht, p 249
94. Carbó-Dorca R, Besalú E, Gironés X (1999) Extended Density Functions. Adv Quantum Chem, in press
95. Carbó-Dorca R (1997) Fuzzy sets and Boolean Tagged Sets. J Math Chem 22:143-147
96. Carbó-Dorca R (1998) Fuzzy sets and boolean tagged sets; vector semispaces and convex sets; quantum similarity mesures and ASA density functions; diagonal vector spaces and quantum chemistry. In: Carbó-Dorca R, Mezey PG (eds) Advances in Molecular Similarity. JAI Press, Greenwich, Vol 2, pp 43-72
97. Carbó-Dorca R (1998) Tagged Sets, Convex Sets and Quantum Similarity Measures. J Math Chem 23:353-364
98. Carbó R, Besalú E, Amat L, Fradera X (1996) On molecular quantum similarity measures (QMSM) and indices (QMSI). J Math Chem 19:47-56
99. Robert D, Carbó-Dorca R (1998) A formal comparison between molecular quantum similarity indices. J Chem Inf Comput Sci 38:469-475
100. Besalú E, Carbó R, Mestres J, Solà M (1995) Foundations and Recent Developments on Quantum Molecular Similarity. Top Curr Chem 173:31-62
101. Arsenin VY (1968) Basic Equations and Special Functions of Mathematical Physics. Iliffe Books, London
102. Matsuoka O (1973) Int J Quantum Chem 7:365-381
103. Bethe AH, Salpeter EE (1957) Quantum mechanics of one- and two-electron Systems. Springer-Verlag, Berlin

104. Dunlap BI, Connnolly JWD, Sabin JR (1979) On some approximations in applications of some X_α theory. J Chem Phys 71:3396-3402

105. Mestres J, Solà M, Duran M, Carbó R (1994) On the calculation of ab initio quantum molecular similarities for large systems: fitting the electron density. J Comput Chem 15:1113-1120

106. Cioslowski J, Piskorz P, Rez P (1997) Accurate analytical representations of the core electron densities of the elements 3 through 118. J Chem Phys 106:3607-3612

107. Constans P, Carbó R (1995) Atomic shell approximation: electron density fitting algorithm restricting coefficients to positive values J Chem Inf Comput Sci 35:1046-1053

108. Constans P, Amat L, Fradera X, Carbó-Dorca R (1996) Quantum molecular similarity measures (QMSM) and the atomic shell approximation (ASA). In: Carbó-Dorca R, Mezey PG (eds) Advances in Molecular Similarity. JAI Press, Greenwich, Vol 1, pp 187-211

109. Amat L, Carbó-Dorca R (1997) Quantum similarity measures under atomic shell approximation: first order density fitting using elementary Jacobi rotations. J Comput Chem 18:2023-2039

110. Amat L, Carbó-Dorca R (1999) Fitted electronic density functions from H to Rn for use in quantum Similarity measures: Cis-diamminedichloroplatinum(II) complex as an application example. J Comput Chem 20:911-920

111. Ruedenberg K, Schwarz WHE (1990) Nonspherical atomic ground-state densities and chemical deformation densities from x-ray scattering. J Chem Phys 42:4956-4969

112. Coppens P (1992) In: International Tables for Crystallography. Kluwer, Amsterdam, Vol B, p 10

113. Coppens P, Becker (1992) In: International Tables for Crystallography. Kluwer, Amsterdam, Vol C, p 628

114. Ruedenberg K, Raffenetti RC, Bardon D (1973) Energy, structure and reactivity. Proceedings of the 1972 Boulder Conference on Theoretical Chemistry. Wiley, New York, p 164

115. Schmidt MW, Ruedenberg K (1979) Effective convergence to complete orbital bases and to the atomic Hartree-Fock limit through systematic sequences of Gaussian primitives. J Chem Phys 71:3951-3962

116. Feller DF, Ruedenberg K (1979) Systematic approach to extended even-tempered orbital bases for atomic and molecular calculations. Theor Chim Acta 52:231-251

117. Jacobi CGJ (1846) Über ein leichtes Verfahren, die in der Theorie der Säkularstörungen vorkommenden Gleichungen numerisch aufzulösen. J Reine Angew Math (Crelle's Journal) 30:51-94

118. Wilkinson JH, Reinsch C (1971) Linear algebra. Springer-Verlag, Berlin, pp 202-211

119. Pierre DA (1969) Optimization theory with applications. Wiley, New York

120. Carbó-Dorca R, Amat L, Besalú E, Gironés X, Robert D (1999) Quantum molecular similarity: theory and applications to the evaluation of molecular properties, biological activities and toxicity. In: Carbó-Dorca R, Mezey PG (eds) The Fundamentals of Molecular Similarity. Kluwer, New York, in press

121. McLean AD, Chandler GS (1980) Contracted gaussian basis sets for molecular calculations. I. Second row atoms, Z=11-18. J Chem Phys 72:5639-5648

122. Krishnan B, Binkley JS, Seeger R, Pople JA (1980) Self-consistent orbital methods. XX. A basis set for correlated wave functions. J Chem Phys 72:650-654

123. These coefficients and exponents can be downloaded from: http://iqc.udg.es/cat/similarity/ASA/func6311.html

124. Gironés X, Amat L, Carbó-Dorca R (1998) A comparative study of isodensity surfaces using "ab initio" and ASA density functions. J Mol Graph Model 16:190-196

125. Atai A, Tomioka N, Yamada M, Inoue A, Kato Y (1993) Molecular superposition for rational drug design. In: Kubinyi H (ed) 3D QSAR in drug design. ESCOM, Leiden, pp 200-225

126. Nyburg SC (1974) Some uses of a best molecular fit routine. Acta Cryst B30:251-253

127. Martin YC (1992) 3D database searching in drug design. J Med Chem 35:2145-2154

128. Constans P, Amat L, Carbó-Dorca R (1997) Toward a global maximization of the molecular similarity function: superposition of two molecules. J Comput Chem 18:826-846

129. See, for example: Bayada DM, Simpson RW, Johnson AP, Laurenco C (1992) An algorithm for the multiple common subgraph problem. J Chem Inf Comput Sci 32:680-685

130. Gavuzzo E, Pagliuca S, Pavel V, Quagliata C (1972) Generation and best fitting of molecular models. Acta Cryst B28:1968-1969

131. McLachlan AD (1972) A mathematical procedure for superimposing atomic coordinates of proteins. Acta Cryst A28:656-657

132. Gerber PR, Muller K (1987) Superimposing several sets of atomic coordinates. Acta Cryst A41:426-428

133. Redington PK (1992) Molfit: A computer program for molecular superposition. Comput Chem 16:217-222

134. Kearsley SK, Smith GM (1990) An alternative method for the alignment of molecular structures: maximizing electrostatic and steric overlap. Tetrahedron Comput Method 3:615-633

135. Manaut M, Sanz F, Jose J, Milesi M (1991) Automatic search for maximum similarity between molecular electrostatic potential distributions. J Comput-Aided Mol Design 5:371-380

136. Clark M, Cramer III RD, Jones DM, Patterson DE, Simeroth PE (1990) Comparative molecular field analysis (CoMFA). 2. Towards its use with 3D-structural databases. Tetrahedron Comput Method 3:47-59

137. Good AC, Hodgkin EE, Richards WG (1992) Utilization of Gaussian functions for the rapid evaluation of molecular similarity. J Chem Inf Comput Sci 32:188-191

138. Mestres J, Rohrer DC, Maggiora GM (1997) MIMIC: A molecular-field matching program. exploiting applicability of molecular similarity approaches. J Comput Chem 18:934-954

139. Parretti MF, Kroemer RT, Rothman JH, Richards WG (1997) Alignment of molecules by the Monte Carlo optimization of molecular similarity indices. J Comput Chem 18:1344-1353

140. McMahon AJ, King PM (1997) Optimization of Carbó molecular similarity index using gradient methods. J Comput Chem 18:151-158

141. Dean PM, Callow P, Chau PL (1988) Molecular recognition: blind-searching for regions of strong structural match on the surfaces of two dissimilar molecules. J Mol Graph 6:28-34

Chapter 3

142. Hansch C, Fujita T (1964) ρ-σ-π Analysis. A method for correlation of biological activity and chemical structure. J Am Chem Soc 86:1616-1626

143. Martin YC (1978) Quantitative Drug Design. A Critical Introduction. Marcel Dekker, New York

144. Cramer III RD, Patterson DE, Bunce JD (1988) Comparative Molecular Field Analysis (CoMFA). 1. Effect of Shape on Binding of Steroids to Carrier Proteins. J Am Chem Soc 110:5959-5967

145. Good AC, So S-S, Richards WG (1993) Structure-Activity Relationships from Molecular Similarity Matrices. J Med Chem 36:433-438

146. Jain AN, Koile K, Chapman D (1994) Compass: Predicting Biological Activities from Molecular Surface Properties. Performance Comparisons on a Steroid Benchmark. J Med Chem 37:2315-2327

147. Klebe G, Abraham U, Mietzner T (1994) Molecular similarity indices in a comparative analysis (CoMSIA) of drug molecules to correlate and predict their biological activity. J Med Chem 37:4130-4146

148. Silverman BD, Platt DE (1996) Comparative molecular moment analysis (CoMMA): 3D-QSAR without molecular superposition. J Med Chem 39:2129-2140

149. Wold S, Johansson E, Cocchi M. (1993) PLS-Partial Least-Squares Projections to Latent Structures. In: Kubinyi H (ed) 3D QSAR in Drug Design. ESCOM, Leiden, pp 523-550

150. Tetko IV, Luik AI, Poda GI (1993) Application of Neural Networks in Structure-Activity Relationships of a Small Number of Molecules. J Med Chem 36:811-814

151. Ajay A (1993) A Unified Framework for Using Neural Networks to Build QSARs. J Med Chem 36:3565-3571

152. Stuper AJ, Jurs C (1975) Classification of Psychotropic Drugsas Sedatives or Tranquilizers using Pattern Recognition Techniques. J Am Chem Soc 97:182-187

153. Kowalski BR, Bender CF (1973) Pattern Recognition. II. Linear and Nonlinear Methods for Displaying Chemical Data. J Am Chem Soc 95:686-693

154. McFarland JW, Gans DJ (1987) Cluster Significance Analysis contrasted with three other quantitative structure-activity relationship models. J Med Chem 30:46-49

155. Moriguchi I, Hirono S, Liu Q, Nakagome I (1992) Fuzzy Adaptive Least Squares and its application to Structure-Activity Studies. Quant Struct-Act Relat 11:325-331

156. Carbó R, Besalú E, Amat L, Fradera X (1995) Quantum molecular similarity measures (MQSM) as a natural way leading towards a theoretical foundation of quantitative structure-properties relationships (QSPR). J Math Chem 18:237-246

157. Montogmery DC, Peck EA (1992) Introduction to linear regression analysis. Wiley, New York

158. Amat L, Carbó-Dorca R, Ponec R (1999) Simple linear QSAR models based on Quantum Similarity Measures. J Med Chem 42:5169-5180

159. Zupan J, Gasteiger J (1993) Neural Networks for Chemists. VCH, Weinheim

160. Allen DM (1974) The relationship between variable selection and data augmentation and a method for prediction. Technometrics 16:125-127

161. Wold S, Eriksson L (1995) Statistical validation of QSAR results. In: Van de Waterbeemd H (ed) Chemometric methods in molecular design. VCH, New York, Vol 2, pp 309-318

162. Kier LB, Hall LH (1974) Molecular Connectivity and Drug Research. Academic Press, New York

163. Hall LH, Kier LB (1991) The Molecular Connectivity Chi Indexes and Kappa Shape Indexes in Structure-Poperty Modelling. In: Lipkowitz KB, Boyd DB (eds) Reviews in Computational Chemistry II. VCH, New York, pp 367-422

164. Mihalic Z, Trinajstic N (1992) A Graph-Theoretical Approach to Structure-Property Relationships. J Chem Educ 69:701-712

165. Cho SJ, Tropsha A (1995) Cross-Validated R2-Guided Region Selection for Comparative Molecular Field Analysis: A Simple Method to Achieve Consistent Results. J Med Chem 38:1060-1066

166. Kroemer TR, Hecht P (1995) Replacement of Steric 6-12 Potential-Derived Interaction energies by Atom-Based Indicator Variables in CoMFA Leads to Models of Higher Consistency. J Comput-Aided Mol Design 9:205-212

167. Klebe G, Abraham U (1993) On the prediction of Binding Properties of Drug Molecules by Comparative Molecular Field Analysis. J Med Chem 36:70-80

168. Sulea T, Oprea TI, Muresan S, Chan SL (1997) A Different Method for Steric Field Evaluation in CoMFA Improves Model Robustness. J Chem Inf Comput Sci 37:1162-1170

169. Folkers G, Merz A, Rognan D (1993) CoMFA: Scope and Limitations. In: Kubinyi H (ed) 3D QSAR in Drug Design. ESCOM, Leiden, pp 583-618

170. Cramer III RD, DePriest SA, Patterson A, Hecht P (1993) The Developing of Comparative Molecular Field Analysis. In: Kubinyi H (ed) 3D QSAR in Drug Design. ESCOM, Leiden, pp 443-485

171. Simon Z (1992) Comparative molecular field analysis. Critical comments. Rev Roum Chem 37:323-325

Chapter 4

172. Van de Waterbeemd H (1996) Chemometric Methods used in Drug Discovery. In: Van de Waterbeemd H (ed) Structure-Property Correlations in Drug Design. Academic Press, San Diego

173. Cuadras CM, Fortiana J (2000) The Importance of Geometry in Multivariate Analysis and some Applications. In: Rao CR, Szekely GJ (eds) Statistics for the 21st Century, Marcel Dekker, New York, pp 93-108

174. Torgerson WS (1952) Multidimensional scaling: I. Theory and method. Psychometrika 17:401-419

175. Richardson MW (1938) Psychological Bulletin 35:659-660

176. Eckart C, Young G (1936) The approximation of one matrix by another of lower rank. Psychometrika 1:211-218

177. Young G, Householder AS (1938) Discussion of a set of points in terms of their mutual distances. Psychometrika 3:19-22

178. De Leeuw J, Heiser W (1982) Theory of multidimensional scaling. In: Krishnaiah PR, Kanal LN (eds) Handbook of Statistics, Vol 2. North Holland, Amsterdam, pp 285-316

179. Mardia KV, Kent JT, Bibby JM (1979) Multivariate Analysis. Academic Press, London

180. Gower JC, Legendre P (1986) Metric and Euclidean properties of dissimilarity coefficients. J Class 3:5-48

181. Cuadras CM, Arenas C (1990) A distance based regression model for prediction with mixed data. Commun Statist Theor Meth 19:2261-2279

182. Cuadras CM, Arenas C, Fortiana J (1996) Some computational aspects of a distance-based model for prediction. Commun Statist Simula 25:593-609

183. Amat L, Robert D, Besalú E, Carbó-Dorca R (1998) Molecular Quantum Similarity Measures Tuned QSAR: An Antitumoral Family Validation Study. J Chem Inf Comput Sci 38:624-631

184. Demidovich BP, Maron IA (1981) Computational Mathematics. Mir Publishers, Moscow

185. Pierre DA (1969) Optimization Theory with Applications. John Wiley, New York

186. Carbó R, Besalú E (1994) Definition, mathematical examples and quantum chemical applications of nested summation symbols and logical Kronecker deltas. Computers Chem 18:117-126

187. Carbó R, Besalú E (1995) Definition and quantum chemical applications of nested summation symbols and logical functions: Pedagogical artificial intelligence devices for formulae writing, sequential programming and automatic parallel implementation. J Math Chem 18:37-72

188. Hadjipavlou-Litina D, Hansch C (1994) Quantitative structure-activity relationships of the benzodiazepines. A review and reevaluation. Chem Rev 94:1483-1505

189. Dewar MJS, Zoebisch EG, Healy EF, Stewart JJP (1985) AM1: A new general purpose quantum chemical molecular model. J Am Chem Soc 107:3902-3909

190. Kaiser KLE (ed) (1987) QSAR in environmental toxicology. Reidel Publishing Company, Dordrecht

191. Hutzinger O (ed) (1989) Handbook of environmental chemistry. Springer-Verlag, Berlin

192. Rekker RF (1977) The hydrophobic fragmental constants. Its derivation and application. A means of the characterization membrane systems. In: Nauta WT, Rekker RF (eds) Pharmacochemistry Library. Elsevier, New York, Vol 1

193. Amat L, Carbó-Dorca R, Ponec R (1998) Molecular quantum similarity measures as an alternative to log P values in QSAR studies. J Comput Chem 19:1575-1583

194. Verhaar HJM, Mulder W, Hermens JLM, Rorije E, Langenberg JH, Peijnenburg WJGM, Sabljic A, Güsten H, Eriksson L, Sjöström M, Müller M, Hansen B, Nouwen J, Karcher W (1995) Overview of Structure-Activity Relationships for Environmental Endpoints. Part 1: General Outline and Procedure. Report of the EU-DG-XII Project QSAR for Predicting Fate and Effects of Chemicals in the Environment. (Contract #EV5V-CT92-0211)

195. Urrestarazu E, Vaes WHJ, Verhaar HJM, Hermens JLM (1998) Quantitative Structure-Activity Relationships for the Aquatic Toxicity of Polar and Nonpolar Narcotic Pollutants. J Chem Inf Comput Sci 38:845-852

196. AMPAC 6.0, 1994 Semichem, 7128 Summit, Shawnee, KS 66216 D.A

197. Cattell RB (1966) the scree test for the number of factors. Multivariate Behavioral Research 9:331-341

198. Oxender DL, Fox CF (eds) (1987) Protein Engineering. Alan R. Liss, New York, pp 221-224

199. Leatherbarrow RJ, Fersht AJ (1986) Protein Engineering. Protein Eng 1:7-16

200. Hellberg S, Sjostrom M, Skagerberg B, Wold S (1987) Peptide quantitative structure-activity relationships, a multivariate approach. J Med Chem 30:1126-1135

201. Kato A, Yutani K (1988) Correlation of surface properties with conformational stabilities of wild-type and six mutant tryptophan synthase alpha-subunits substituted at the same position. Protein Eng 2:153-156

202. Lee C, Levitt M (1991) Accurate prediction of the stability and activity effects of site-directed mutagenesis on a protein core. Nature 352:448-451

203. Collantes ER, Dunn WJ III (1995) Amino acid side-chain descriptors for quantitative structure-activity relationship studies of peptide analogues. J Med Chem 38:2705-2713

204. Zbilut JP, Giuliani A, Webber CL Jr, Colosimo A (1998) Recurrence quantification analysis in structure-function relationships of proteins: an overview of a general methodology applied to the case of TEM-1 beta-lactamase. Protein Eng 11:87-93

205. Hutchison CA III, Phillips S, Edgell MH, Gillam S, Jahnke P, Smith M (1978) Mutagenesis at a specific position in a DNA sequence. J Biol Chem 253:6551-6560

206. DeSantis G, Berglund P, Stabile MR, Gold M, Jones JB (1998) Site-directed mutagenesis combined with chemical modification as a strategy for altering the specificity of the S1 and S1' pockets of subtilisin Bacillus lentus. Biochem 37:5968-5973

207. Mei HC, Liaw YC, Li YC, Wang DC, Takagi H, Tsai YC (1998) Engineering subtilisin YaB: restriction of substrate specificity by the substitution of Gly124 and Gly151 with Ala. Protein Eng 11:109-117

208. Narinx E, Baise E, Gerday C (1997) Subtilisin from psychophilic antarctic bacteria: characterization and site-directed mutagenesis of residues possibly ivolved in the adaptation to cold. Protein Eng 10:1271-1279

209. Takagi H, Ohtsu I, Nakamori S (1997) Construction of novel subtilisin E with high specificity, activity and productivity through multiple amino acid substitutions. Protein Eng 10:985-989

210. Stauffer CE, Etson D (1969) The effect of subtilisin activity on oxidizing a methionine residue. J Biol Chem 244:5333-5338

211. Estell DA, Graycar TP, Wells JA (1985) Engineering an enzyme by site-directed mutagenesis to be resistant to chemical oxidation. J Biol Chem 260:6518-6521

212. Wells JA, Vasser M, Powers DB (1985) Cassette mutagenesis: an efficient method for generation of multiple mutations at defined sites. Gene 34:315-323

213. PC Spartan Plus, Wavefunction Inc., Irvine, CA 92612 USA

214. Damborský J (1998) Quantitative structure-function and structure-stability relationships of purposely modified proteins. Protein Eng 11:21-30

215. Nakai K, Kidera A, Kanehisa M (1988) Cluster analysis of amino acid indices for prediction of protein structure and function. Protein Eng 2:93-100

216. Tomii K, Kanehisa M (1996) Analysis of amino acid indices and mutation matrices for sequence comparison and structure prediction of proteins. Protein Eng 9:27-36

217. Kawashima S, Ogata H, Kanehisa M (1999) AAindex: amino acid index database. Nucleic Acids Res 27:368-369

218. Sneath PH (1966) Relations between chemical structure and biological activity in peptides. J Theor Biol 12:157-195

219. Oobatake M, Ooi T (1977) An analysis of non-bonded energy of proteins. J Theor Biol 67:567-584

220. Wells JA, Powers DB, Bott RR (1987) In: Oxender DL, Fox CF (eds) Protein engineering. Alan R Liss, New York, pp 279-287

221. Robert D, Gironés X, Carbó-Dorca R (1999) Quantification of the influence of single-point mutations on Haloalkane dehalogenase activity: a quantum similarity study. J Chem Inf Comput Sci. in press

222. Carbó-Dorca R, Amat L, Besalú E, Gironés X, Robert D (1999) Quantum molecular similarity: theory and applications to the evaluation of molecular properties, biological activities and toxicity. In: Carbó-Dorca R, Mezey PG (eds) The Fundamentals of Molecular Similarity. Kluwer, New York, in press

Chapter 5

223. Kubinyi H (1993) QSAR: Hansch Analysis and Related Approaches. VCH, Weinheim

224. Fujita IT, Iwasa J, Hansch C (1964) A new substituent constant, π, derived from partition coefficients. J Am Chem Soc 86:5175-5180

225. Rekker RF (1977) The hydrophobic fragmental constants. Its derivation and applications. A means of characterizing membrane systems. Elsevier, New York

226. Hansch C, Leo A (1979) Substituent constants for correlation analysis in chemistry and biology. Wiley, New York

227. Klopman G, Iroff L (1981) Calculation of partition coefficients by the charge density method. J Comput Chem 2:157-160

228. Klopman G, Namboodiri K, Schochet M (1985) Simple method of computing the partition coefficent. J Comput Chem 1:28-38

229. Ghose AK, Crippen GM (1986) Atomic physicochemical parameters for three-dimensional structure-directed quantitative structure-activity relationships. I. Partition coefficients as a measure of hydrophobicity. J Comput Chem 7:565-577

230. Koehler MG, Grigoras S, Dunn WJ III (1988) The relationship between chemical structure and the logarithm of the partition coefficient. Quant Struct-Act Relat 7:150-159

231. Bodor N, Gabanyi Z, Wong C-K (1989) A new method for the estimation of partition coefficient. J Am Chem Soc 111:3783-3786

232. Richards NGJ, Williams PB, Tute MS (1991) Empirical methods for computing molecular partition coefficients. I. Upon the need to model the specific hydration of polar groups in fragment-based approaches. Int J Quantum Chem, Quantum Biology Symposium 18:299-316

233. Richards NGJ, Williams PB, Tute MS (1992) Empirical methods of computing molecular partition coefficients: II. Inclusion of conformational flexibility within fragment-based approaches. Int J Quantum Chem 44:219-233

234. Essex JW, Reynolds CA, Richards WG (1992) Theoretical determination of partition coefficients. J Am Chem Soc 114:3634-3639

235. Moriguchi I, Hirono S, Liu Q, Nakagome I, Matsushita Y (1992) Simple method of calculating octanol/water partition coefficient. Chem Pharm Bull 40:127-130

236. Moriguchi I, Hirono S, Nakagome I, Hirano H (1994) Comparison of reliability of log P values for drugs calculated by several methods. Chem Pharm Bull 42:976-978

237. Amat L, Carbó-Dorca R, Ponec R (1998) Molecular quantum similarity measures as an alternative to log P values in QSAR studies. J Comput Chem 19:1575-1583

238. Ponec R, Amat L, Carbó-Dorca R (1999) Molecular basis of quantitative structure-properties relationships (QSPR): A quantum similarity approach. J Comput-Aided Mol Design 13:259-270

239. Amat L, Carbó-Dorca R, Ponec R (1999) Simple linear QSAR models based on Quantum Similarity Measures. J Med Chem 42:5169-5180

240. Hansch C, Kim D, Leo AJ, Novellino E, Silipo C, Vittoria A (1989) Toward a quantitative comparative toxicology of organic compounds. CRC Crit Rev Tox 19:185-226

241. Dewar MJS, Zoebisch EG, Healy EF, Stewart JJP (1985) AM1: A new general purpose quantum chemical molecular model. J Am Chem Soc 107:3902-3909

242. AMPAC 6.0, 1994 Semichem, 7128 Summit, Shawnee, KS 66216 D.A.

243. Hammett LP (1938) Linear free energy relationships in rate and equilibria phenomena. Trans Faraday Soc 34:156

244. Ponec R, Amat L, Carbó-Dorca R (1999) Quantum similarity approach to LFER: substituent and solvent effects on the acidities of carboxylic acids. J Phys Org Chem 12:447-454

245. Lesher GY, Froelich EJ, Gruett MD, Bailey JH, Brundage RP (1962) 1,8-Naphthyridine derivatives. A new class of chemotherapeutic agents. J Med Chem 5:1063-1065

246. Koga H, Ito A, Murayama S, Suzue S, Irikura T (1980) Structure-activity relationships of antibacterial 6,7 and 7,8-disubstituted 1-alkyl-1,4-dihydro-4-oxoquinoline-3-carboxylic acids. J Med Chem 23:1358-1363

247. Albrecht R (1977) Development of antibacterial agents of the nalidixic acid type. Prog Drug Res 21:9-104

248. Andriole VT (ed) (1988) The quinolones. Academic Press, New York

249. Andriole VT (1990) Quinolones. In: Mandell GL, Douglas RG Jr, Bennett JE (eds) Principles and practice of infectious diseases, 3rd Ed. Churchill Livingstone, New York, pp 334-345

250. Cornett JB, Wentland MP (1986) Quinolone antibacterial agents. Annu Rep Med Chem21:139-148

251. Fernandes PB, Chu DTW (1987) Chemotherapeutic agents. Quinolones. Annu Rep Med Chem 22:117-126

252. Fernandes PB, Chu DTW (1988) Quinolone antibacterial agents. Annu Rep Med Chem 23:133-140

253. Heck JV (1989) Antibacterial agents. Annu Rep Med Chem 24:101-110

254. White DR, Davenport LC (1990) Antibacterial agents. Annu Rep Med Chem 25:109-118

255. Brighty KE, McGuirk PR (1991) Antibacterial agents. Annu Rep Med Chem 26:123-132

256. Frigola J, Parés J, Corbera J, Vañó D, Mercè R, Torrens A, Más J, Valentí E (1993) 7-azetidinylquinolones as antibacterial agents. Synthesis and structure-activity relationships. J Med Chem 36:801-810

257. Carbó R, Besalú E (1994) Definition, mathematical examples and quantum chemical applications of nested summation symbols and logical Kronecker deltas. Computers Chem 18:117-126

258. Carbó R, Besalú E (1995) Definition and quantum chemical applications of nested summation symbols and logical functions: Pedagogical artificial intelligence devices for formulae writing, sequential programming and automatic parallel implementation. J Math Chem 18:37-72

259. Amat L, Carbó-Dorca R (1999) Fitted electronic density functions from H to Rn for use in quantum similarity measures: *cis*-diammine-dichloroplatinum(II) complex as an application example. J Comput Chem 20:911-920

260. Goodman Gilman A, Rall TW, Nies AS, Taylor P (1990) The pharmacological basis of therapeutics. Pergamon Press, New York, p 1024

Chapter 6

261. Roothaan CCJ (1951) New developments in molecular orbital theory Revs Mod Phys 23 69-89

262. Carbó R, Domingo L (1987) LCAO-MO similarity measures and taxonomy Int J Quantum Chem 23 517-545

263. Gironés X, Amat L, Carbó-Dorca R (1999) Using molecular quantum similarity measures as descriptors in quantitative structure-toxicity relationships. SAR QSAR Environ Res, in press

264. Carbó-Dorca R, Amat L, Besalú E, Gironés X, Robert D (1999) Quantum mechanical origin of QSAR: theory and applications. J Mol Struct (Theochem), in press

265. Dewar MJS, Zoebisch EG, Healy EF, Stewart JJP (1985) AM1: A new general purpose quantum chemical molecular model. J Am Chem Soc 107:3902-3909

266. AMPAC 6.01 (1994) Semichem, 7128 Summit, Schawnee, KS 66216DA

267. Frisch MJ, Trucks GW, Schlegel HB, Gill PMW, Johnson BG, Robb MA, Cheeseman JR, Keith T, Petersson GA, Montgomery JA, Raghavachari K, Al-Laham MA, Zakrzewski VG, Ortiz JV, Foresman, JB, Cioslowski J, Stefanov BB, Nanayakkara A, Challacombe M, Peng CY, Ayala PY, Chen W, Wong MW, Andres JL, Replogle ES, Gomperts R, Martin RL, Fox DJ, Binkley JS, Defrees DJ, Baker J, Stewart JP, Head-Gordon M, Gonzalez C, Pople JA (1995) Gaussian-94, (Revision E.2) Gaussian, Inc. Pittsburgh PA

268. Vinter V (1970) Germination and outgrowth: effect of inhibitors. J Appl Bacteriol 33:50-59

269. Yasuda-Yasaki Y, Namiki-Kanie S, Hachisuka Y (1978) Inhibition of germination of Bacillus subtilis spores by alcohols. In: Chambliss G, Vary JC (eds) Spores VII. American Society of Microbiology, Washington, pp 113-116

270. Yasuda-Yataki Y, Nimihi-Kanie S, Hachisuka Y (1978) Inhibition of *Bacillus subtillis* spore germination by various hydrophobic compounds: demonstration of hydrophobic character of the L-alanine receptor site. J Bacteriol 136:484-490

271. Roth JS (1954) Cancer Res 2:346-350

272. Cooley NR, Keltner, JM Jr, Forester J (1973) Polychlorinated biphenyls, aroclors 1248 and 1260: effect on and accumulation by *Tetrahymena pyriformis*. J Protozool 20:443-445

273. Apostol S (1973) Environ Res 6:365-372

274. Dive D, LeClerc H (1975) Prog Water Technol 7:67-72

275. Hill DL (1972) The biochemistry and physiology of Tetrahymena. Academic Press, New York

276. Schultz TW, Lin DT, Wilke TS, Arnold LM (1990) Quantitative structure-activity relationships for the tetrahymena pyriformis population growth endpoint: a mechanisms of action approach. In: Karcher W, Devillers J (eds) Practical Applications of Quantitative Structure-Activity Relationships (QSAR) in Environmental Chemistry and Toxicology. Kluwer Academic Publishers, Dordrecht

277. Urrestarazu E, Vaes WHJ, Verhaar HJM, Hermens JLM (1998) Quantitative structure-activity of the aquatic toxicity of polar and nonpolar aquatic pollutants. J Chem Inf Comput Sci 38:845-852

278. Leibman KC, Ortiz E (1971) Pharmacologist 13:223

279. Leibman KC (1971) Chem Biol Interact 3:289

280. Wilkinson CF, Hetnarski K, Yellin TO (1972) Imidazole derivatives –a new class of microsomal enzyme inhibitors. Biochem Pharmac 21:3187-3192

281. Leibman KC, Ortiz E (1973) Metyrapone and other modifiers of microsomal drug metabolism. Drug Metab Dispos 1:184-190

282. Leibman KC, Ortiz E (1973) New potent modifiers of liver microsomal drug metabolism. Drug Metab Dispos 1:775-779

283. Wilkinson CF, Hetnarski K, Hicks LJ (1975) Pestic Biochem Physiol

284. Wilkinson CF, Hetnarski K, Cantwell P, Di Carlo F (1974) Structure-activity relationships in the effects of 1-alkylimidazoles on microsomal oxidation in vitro and in vivo. J Biochem Pharmacol 23:2377-2386

285. Fujita T, Iwasa J, Hansch C (1964) A new substituent constant, π, derived from partition coefficients. J Am Chem Soc 86:5175-5180

286. Gaudette LE, Brodie BB (1959) Biochem Pharmac 2:89

287. Martin YC, Hansch C (1971) Influence of hydrophobic character on the relative rate of oxidation of drugs by rat liver microsomes. J Med Chem 14:777-779

288. Hansch C (1972) Quantitative relationships between lipophilic character and drug metabolism. Drug Metab Rev 1:1-14

Chapter 7

289. Robert D, Carbó-Dorca R (1998) On the extension of quantum similarity to atomic nuclei: Nuclear quantum similarity. J Math Chem 23:327-351

290. Pan X-W, Feng DH, Vallieres M (eds) (1997) Contemporary Nuclear Shell Models. Springer, New York

291. Friedrich J, Reinhard P-G (1986) Skyrme-force parametrization: Least-squares fit to nuclear ground-state properties. Phys Rev C33:335-351

292. Skyrme THR (1959) The effective nuclear potential. Nucl Phys 9:615-634

293. Vautherin D, Brink DM (1972) Hartree-Fock calculations with Skyrme's interaction. I. Spherical nuclei. Phys Rev C5:626-647

294. Bartel J, Quentin P, Brack M, Guet C, Hakansson H-B (1982) Towards a better parametrisation of Skyrme-like effective forces: a critical study of the SkM force. Nucl Phys A386:79-100

295. Vautherin D (1973) Hartree-Fock calculations with Skyrme's interaction. II. Axially deformed nuclei. Phys Rev C7:296-316

296. Ring P, Schuck P (1980) The Nuclear Many Body Problem. Springer, Berlin

297. Lawson RD (1980) Theory of the Nuclear Shell Model. Oxford, New York.

298. Audi G, Wapstra AH (1995) The 1995 Update to the Atomic Mass Evaluation. Nucl Phys A595:409

299. Weizsäcker CF (1935) Z Phys 96:431

300. Bethe HA, Bacher RF (1936) Revs Mod Phys 8:193

301. Krane KS (1988) Introductory Nuclear Physics. Wiley, New York

302. Robert D, Carbó-Dorca R (1998) Structure-property relationships in nuclei: prediction of the binding energy per nucleon using a quantum similarity approach. Il Nuovo Cimento A111:1311-1321

303. Haustein PE (1988) An overview of the 1986-1987 atomic mass predictions. At Data Nucl Data Tables 39:185-393

Lecture Notes in Chemistry

For information about Vols. 1–32
please contact your bookseller or Springer-Verlag

Editorial Policy

This series aims to report new developments in chemical research and teaching - quickly, informally and at a high level. The type of material considered for publication includes:

1. Preliminary drafts of original papers and monographs

2. Lectures on a new field, or presenting a new angle on a classical field

3. Seminar work-outs

4. Reports of meetings, provided they are
   a) of exceptional interest and

   b) devoted to a single topic.

Texts which are out of print but still in demand may also be considered if they fall within these categories.

The timeliness of a manuscript is more important than its form, which may be unfinished or tentative. Thus, in some instances, proofs may be merely outlined and results presented which have been or will later be published elsewhere. If possible, a subject index should be included. Publication of Lecture Notes is intended as a service to the international chemical community, in that a commercial publisher, Springer-Verlag, can offer a wider distribution to documents which would otherwise have a restricted readership. Once published and copyrighted, they can be documented in the scientific literature.

Manuscripts

Manuscripts should comprise not less than 100 and preferably not more than 500 pages. They are reproduced by a photographic process and therefore must be submitted in camera-ready form according to Springer-Verlag's specifications: technical instructions will be sent on request.

The text area should take care of the page length and width (12.2 x 19.3 cm when you use a 10 point font size, 15.3 x 24.2 cm for a 12 point font size).

Authors receive 50 free copies and are free to use the material in other publications.

Manuscripts should be sent to one of the editors or directly to Springer-Verlag, Heidelberg.